DISCARDED SCIENCE

JOHN GRANT

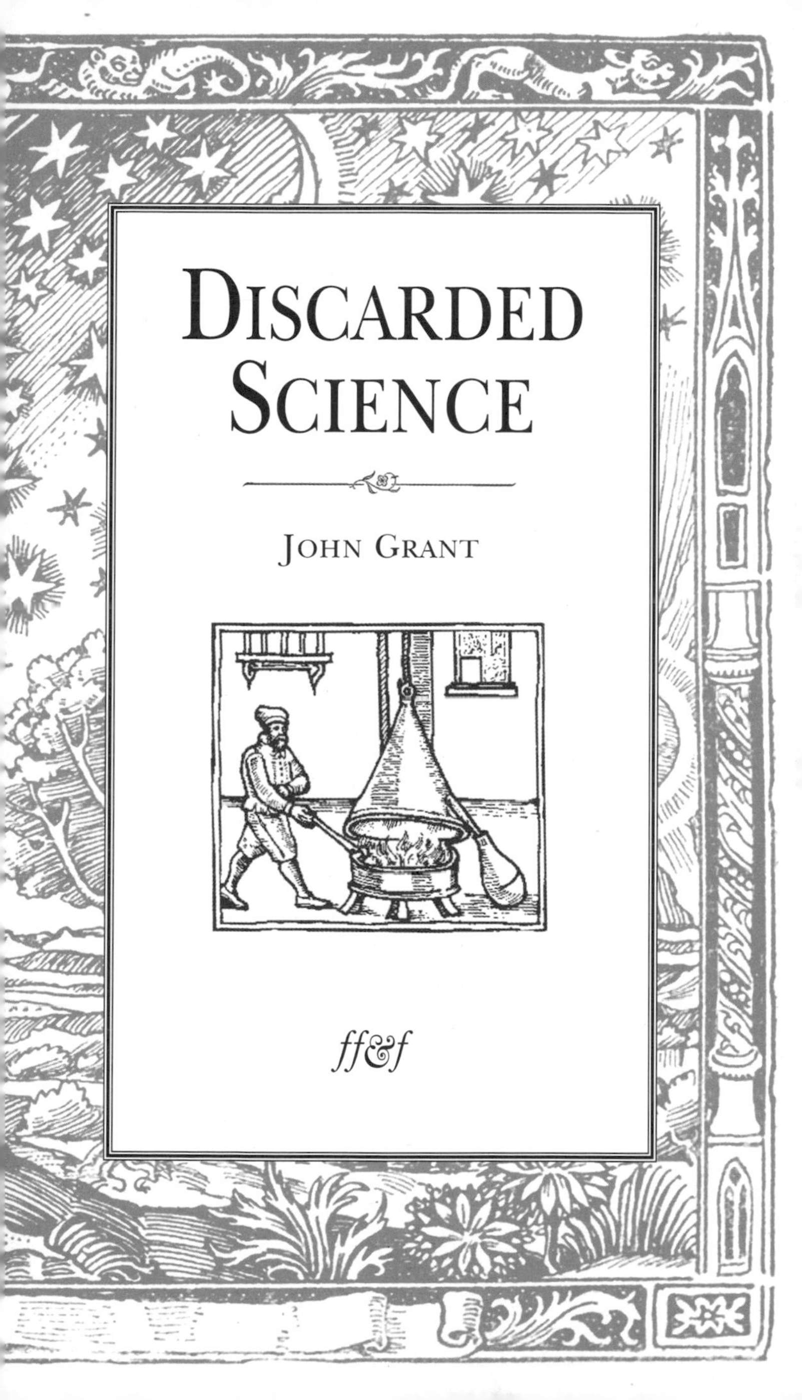

DISCARDED SCIENCE

JOHN GRANT

ff&f

This one's for The Spammers: Randy M. Dannenfelser, Bob Eggleton, Gregory Frost, Neil Greenberg, Jael, Stuart Jaffe, Karl Kofoed, Todd Lockwood, Aaron McLellan, Lynn Perkins, Tim Sullivan and Greg Uchrin.

ACKNOWLEDGEMENTS

John Baez, Cameron Brown, Malcolm Couch, Bill DeSmedt, Neil Greenberg, Ian Johnson, Lynn Perkins, The Spammers, Pamela D. Scoville, and the Union of Concerned Scientists.

Some parts of this book have been drawn, with very extensive modifications, from my earlier book *A Directory of Discarded Ideas* (1981). The section on the Tunguska Event is drawn, again with extensive revision, from my own essay in *A Directory of Possibilities* (1981), which I co-edited with Colin Wilson.

DISCARDED SCIENCE

Published by Facts, Figures & Fun, an imprint of
AAPPL Artists' and Photographers' Press Ltd.
Church Farm House, Wisley, Surrey GU23 6QL
info@ffnf.co.uk www.ffnf.co.uk
info@aappl.com www.aappl.com

Sales and Distribution
UK and export: Turnaround Publisher Services Ltd.
orders@turnaround-uk.com
USA and Canada: Sterling Publishing Inc. sales@sterlingpub.com
Australia & New Zealand: Peribo Pty. peribomec@bigpond.com
South Africa: Trinity Books. trinity@iafrica.com

A catalogue record for this book is available from the British Library.

ISBN 13: 9781904332497
ISBN 10: 1904332498

Cover Design: Stefan Nekuda office@nekuda.at
Content Design: Malcolm Couch mal.couch@blueyonder.co.uk

Printed in China by Imago Publishing info@imago.co.uk

CONTENTS

Atlas bears the world on his shoulders.
From William Cuningham's *The Cosmographical Glasse*, 1554.

INTRODUCTION

All Manner of Discarded Science

Nothing but experience could evince the frequency of false information, or enable any man to conceive that so many groundless reports should be propagated, as every man of eminence may hear of himself. Some men relate what they think, as what they know; some men of confused memory and habitual inaccuracy, ascribe to one man what belongs to another; and some talk on, without thought or care, A few men are sufficient to broach falsehoods, which are afterwards innocently diffused by successive relaters.

Samuel Johnson, as reported by
James Boswell, *Life of Johnson*, 1778

In 1938 the young Italian sculptor Francesco Cremonse buried in France most of a "classical" statue of Venus. When this statue was "discovered" shortly afterwards, the experts enthused over it and the French government classified it as a part of the national heritage. Cremonse promptly confessed responsibility for creating the sculpture, and was of course disbelieved – who was he to argue with the experts? He was taken seriously only when he produced the missing parts of the statue – its legs, one arm, and its nose. Even so, his outrageous claims might have been swept under the carpet had he not been able to produce the night-club singer who had been his model – complete with legs, arms and nose.

In the field of discarded science one finds, constantly, a similar reluctance on the part of most of us to accept that one or other of our favourite hypotheses is nonsensical. True, some are so overtly misguided that no one sane would take them seriously – for example, there was a Black Forest group, Vegetaria Universa, active in the 1960s, which claimed that the Universe is made entirely of vegetables. But some equally hare-brained ideas are widely shared. Take for example the ancient-astronaut hypotheses of Erich von Däniken (b1935) and others, or the bizarre ideas of the late Immanuel Velikovsky (1895–1979).

The motivations for producing and/or believing in spurious hypotheses are several.

Some hypotheses – indeed, some full-fledged theories – are simply **defunct science**: in their day they represented the cutting-edge of theoretical science but they've since been superseded by other versions that accord more accurately with reality, or they've been realized to be totally at odds with reality and been replaced wholesale by something that is at least, so far as we can establish, closer to the truth. As science slowly evolves – as our knowledge of the Universe slowly progresses in the direction of completeness, even if as yet maybe nowhere approaching that ideal closely – ideas and hypotheses from seemingly different disciplines suddenly take on new relationships: can be seen as the myriad pieces of a single, very large jigsaw. Those pieces that simply will not fit, no matter how much we manipulate them, naturally come under greatest scrutiny. Perhaps, just perhaps, getting them to fit will involve a paradigm shift – the removal of all the other pieces of the jigsaw in order to start afresh. More likely, they're from the wrong puzzle and should never have got into this one's box in the first place.

Some jigsaw pieces very obviously do not belong to the puzzle at all. We should in theory look at them closely, of course, just in case our first impressions might have misled us; but the chances of this being so are slender. There have been instances where "manifestly wrong" pieces have turned out to fit into the larger puzzle after all: for a single example, we can think of Alfred Wegener's hypothesis of continental drift, which for decades was ridiculed by Earth scientists. But continental drift is one of the very, very rare exceptions. Most of those misshapen, incongruously coloured hypotheses are rightly rejected on sight as **pseudoscience**: at a cursory glance they might look like science, but they aren't. While in one sense the pseudosciences are astonishingly unimportant – not only do they make no contribution whatsoever to the advance of human knowledge but very often they could impede it – in another, societal sense they can be of very great importance indeed: through their capacity to deceive the unwary, the uneducated or (to be frank) the bigoted and the stupid, they can sway whole societies, and often can cause untold misery.

A subject bedevilled by pseudoscience is race. The racist pseudosciences were in large part born from the **ideological corruption of science**, a further spawner of a plethora of bad science. The ideology concerned can be religious or it can be political. The ideology can even be, on the whole, a benevolent one. There are growing signs at the moment of science being corrupted by the wholly spurious notion that somehow truth can be determined by democratic vote: if polls show that 60% or 70% of the people don't accept that humankind is the product of Darwinian evolution, for example, that's accepted by some as "proof" that Darwinian evolution must be false. More often, though, the ideology concerned is malicious, or at the very least pernicious in its intent towards science. The corruption of science by religious ideology has been such a constant and pervasive theme throughout history that, had I allowed it to, it would have been mentioned on many more pages of this book than it is; in the vast majority of instances, however, I've assumed the reader will understand that religious belief has moulded the social climate responsible for the acceptance of false science. Elsewhere the corruption of science has been deliberate and politically motivated; the examples that spring at once to mind are the Nazi promotion in Germany of "non-Jewish" science, the Stalinist suppression in the USSR of honest genetics in favour of the populist, "peasant" pseudo-genetics propounded by T.D. Lysenko (1898–1976), and the ongoing attempts by the George W. Bush Administration in the US to stifle science that clashes with its neoconservative ideology.

The ideological corruption of science blends naturally into **antiscience**, the emotional rejection of *all* scientific conclusions, often because those conclusions have been reached by "them" and must therefore be antithetical to "us" – i.e., must be false. This motive clearly spurs quite a lot of the religiously rooted corruption of science: it is easier for a demagogic preacher to persuade his congregation that science *in toto* is the instrument of Satan, and therefore to be rejected wholesale, than it is for him to offer any rational, or at least quasi-rational, counter to whichever piece of science it is that he doesn't happen to like. That shouldn't be taken to mean, though, that religion is the root of all evil so far as antiscience is concerned: there are other motivations, such as that strange form of inverse snobbery in which the highly educated are somehow regarded as less well informed than the uneducated. Here antiscience shades into the **conspiracy-theory** worldview whereby the clever are inferior to the stupid and thus must be plotting to the detriment of the latter. Conspiracy theories abound, manifestly, among the adherents of the pseudosciences: the denizens of orthodox science are secretly conspiring to ignore or denigrate *my* theory. This gives the pseudoscientist a false feeling of importance, because the truth almost always is that scientists are ignoring him because they have better things to do with their time – like genuine science – than spell out the reasons why hogwash is hogwash.

Yet another contributor to the corruption of science is fraud. Fraud by the layman is usually described as hoaxing, is most frequently done for profit, and is perhaps best described as deliberate pseudoscience; at least some pseudoscientists are anyway spurred solely by the profit motive – just go browsing on the internet and see how many sham astrologers ask for your money in the first five minutes – and therefore are really fraudsters in all but name. But the more serious issue here is fraud perpetrated by scientists themselves, perhaps in the pursuit of career advancement, perhaps in the hopes of fame and glory, occasionally – as per the faking by Cyril Burt (1883–1971) of his IQ studies – for reasons that are essentially ideological. Scientists like to pretend that fraud within the scientific establishment is rare and anyway rapidly detected, thanks to the scientific process itself – the process of peer review, attempted replication of experiments, and so on – not to mention, of course, the inherent honesty of scientists. Well, perhaps. Serious cases of scientific fraud are sufficiently numerous that whole books have been written about them, and the frequency of such frauds seems to have been steadily increasing. There have already been several spectacular cases since the start of the 21st century.

There is of course overlap between all these categories of motivations for producing or subscribing to bad science. Although I hesitate to pick on Creationism, it is at one and the same time defunct science and a "democratic" pseudoscience, is corrupted by (in this instance religious) ideology, is sustained in large part by antiscience prejudices, is riddled by fraud – particularly in its guise of Intelligent Design, although the Creationist selectivity in the insistence on a literal reading of the Bible is in itself fraudulent (let's ignore the parts about it being okay to rape virgins and so on, and focus instead on the Genesis account of Creation) – and is fuelled to a great extent by the conspiracy theory that "materialists" will do just about anything in their relentless quest to undermine society's morals and deprave us all.

In the wake of the 2004 tsunami that devastated huge areas of the Indian Ocean border and exacted a horrific death toll, science was very swift to identify causes and propose means of ensuring that, in any future such event, the consequences could be at least minimized. The contrasting irrational response was exemplified on the Indonesian island of Aceh. There it was officially proclaimed by Islamic clerics and lawmakers that the tsunami had been caused by the sinfulness of Aceh women, and the Sharia police moved into full overdrive persecuting women for such tsunami-causing crimes as failing to wear headscarves. What proof could the lawmakers offer that it was the shameless women who were responsible for the disaster? Well, none. Did the persecution help mitigate the plight of the disaster's survivors, such as lack of sufficient medical care and homelessness? Well, no. Did it offer a theological camouflage for the sado-sexual pleasure the male persecutors gained from maltreating and humiliating attractive women? There, science might be able to offer an answer.

Similar irrationality falls easily from the lips of influential Christian Fundamentalist demagogues who, to the dismay of their more rational titular coreligionists, dominate the US airwaves – and from no lips is the fall readier than those of the Reverend Pat Robertson (b1930), founder of the Christian Coalition and the Christian Broadcasting Network. In the lead-up to the 2005 Dover, Pennsylvania, trial over the teaching of Intelligent Design in the science classes of the district's schools (see page 188), the electorate of Dover took the opportunity, in the November 2005 election, to oust all eight members of the school board who had supported the scheme. On November 10, on his daily TV show *The 700 Club*, Robertson warned:

> I'd like to say to the good citizens of Dover: if there is a disaster in your area, don't turn to God – you just rejected him from your city. And don't wonder why He hasn't helped you when your problems begin. I'm not saying they will, but if they do, just remember, you just voted God out of your city. And if that's the case, don't ask for His help because he might not be there.

Either (a) this remarkable statement is completely irrational or (b) there is a God, Robertson has a direct line of communication with Him, and God has told him He likes ID. If the relationship is that close, should Dover in future be hit by any unheralded catastrophe, there is an obvious first course of action open to the survivors: sue Pat Robertson. After all, he will have failed to intercede with God to prevent the disaster and thus will be an accomplice to it.

Blaming disasters on God – even pre-emptively, like Robertson – is nothing new. In November 1755 the most destructive earthquake ever to strike the northeastern US hit at Cape Ann, some 50km south of Boston. The Reverend Thomas Prince, of South Church, Boston, knew at once who was to blame: Benjamin Franklin (1706–1790), for having invented the lightning conductor. Before Franklin's scheme of putting pointed metal rods on tall buildings had been universally adopted, God had been able to express His wrath by blasting something with lightning. Now that the presumptuous Franklin had taken that option away from Him, He was having to use earthquakes instead.

Arthur C. Clarke (b1917) formulated a famous Law: "When a distinguished but elderly scientist states that something is possible, he is almost certainly right. When he states that something is impossible, he is very probably wrong." It is not so many years ago that many "distinguished but elderly" scientists told us that spaceflight was impossible. Again, we have G.W.F. Hegel (1770–1831), around 1800, saying that no object in the Solar System remained to be discovered; the first asteroid was found at the beginning of 1801. Auguste Comte (1798–1857) told us in 1835 that we would never know the true natures of the stars, which would forever remain, for us, only useful celestial signposts; within a few decades of his death the spectroscope told us so much about the chemistry – and hence the physics – of the stars that we knew more about them than about the planets of our own Solar System. Thomas Aquinas (1225–1274), in listing those few things which God cannot do, included the construction of a triangle whose interior angles add up to more or less than 180°; yet this is something which we can all do, by the simple expedient of drawing the triangle on a curved surface.

The point of all this is that the "sensible" theories of the respectable are just as likely to be erroneous as are the outpourings of the amateurs. The illustrious are not immune – far from it. W.E. Gladstone (1809–1898) thought the ancient Greeks were colour blind – owing to the lack of "colour-words" in the works of Homer. George Bernard Shaw (1856–1950) had a theory that disease epidemics are due to laundries, because infectious handkerchiefs are sent there.

The way in which credence in spurious claims can build up in the popular mind can be illustrated by the tale of King Tutankamun's curse. "As we all know", this supposed curse devastated the archaeological team responsible for opening up the tomb. In truth, the only one of the principals to die soon after the tomb was opened was Lord Carnarvon (1866–1923); his death came as no surprise at the time, though, because he had been ailing for some while – he had been in fragile health ever since an automobile accident in 1901, and the very reason he became interested in Egyptology was because, after recovering from the immediate effects of the accident, he had taken up the habit of wintering in Egypt. The remaining principals survived on average for a further 24 years after the expedition, reaching an average age of 73. Howard Carter (1874–1939) lasted another 16 years.

And then there are scientists themselves. The important physical chemist Robert Boyle (1627–1691), who around 1662 deduced Boyle's Law concerning the behaviour of gases, in an unrelated field of research suggested it might be a good idea to interview miners to find out if they ever met any demons. Among those who refused to believe meteorites could fall from heaven was Sir Isaac Newton (1642–1727), who made his opinions plain in 1704. He could not envisage any possible source for them, and therefore he declared the idea unfeasible. Francis Bacon (1561–1626), renowned for having derived the Scientific Method, was not entirely the cold rationalist we tend to imagine: another of his notions was that witchcraft could have its origins in the actions of malign spirits.

In May 1872 the great French astronomer Joseph de Lalande (1732–1807) published a long article in *Le Journal de Paris* patiently explaining to his readers all the reasons why thoughts of manned flight in hot-air balloons were the most foolish of pipedreams. On June 5 1783, just 13 months later, the Montgolfier Brothers made their first flight. This was not a good time for technological prediction in Paris. In July 1783, mere weeks after the Montgolfiers' flight, the engineer Claude,

Marquis de Jouffroy d'Abbans (1751–1832), launched a small paddle-wheel steamboat, the *Pyroscaphe*, on the Seine. In response to the successful voyage, the government handed the invention to French Academy of Sciences for evaluation. They replied that the invention was a waste of money and steam-powered water transport a matter not worth pursuing. A few years later, still unable to get any backing, Jouffroy d'Abbans had to flee from the Revolution. Ironically, when in 1803 the US engineer Robert Fulton (1765–1815) lifted steam-powered transport emphatically back into consideration, it was on the Seine that he performed his pivotal test voyages.

The history of science is littered with countless similar examples of scientists proving profoundly *wrong* – especially when working outside their chosen fields. People who have scientific training can be as much amateurs as anyone else when they stray into disciplines where their level of knowledge is not especially higher than that of the lay person. The classic example of this – the one that's almost always cited – is that of the excursion in 1973 of the distinguished mathematician John Taylor (b1931) into the alien field of parapsychology, when he investigated batches of British children who claimed to be able to emulate the much-vaunted "psychic" powers of Uri Geller. Parapsychology is a science insofar as the study of "psychic" claims should be done scientifically. (The term is more generally used to embrace the claims themselves, in which case parapsychology is probably more properly termed a pseudoscience, although even then it doesn't fit the definition very well – which is why there's not a great deal on the subject in this book.) In the scientific study of "paranormal" claims, it's a fundamental requirement, as a safeguard against fraud, to have a good working knowledge of conjuring, since almost all of the effects exhibited by "psychics" are attained in this manner: if an effect can be reproduced using sleight-of-hand, then there's every reason to believe this was how it was done in the first place. Taylor did not have this basic conjuring knowledge – and nor did it seem to occur to him that any of the little darlings involved in his testing might resort to straightforward cheating . . . which of course, being bright kids, many of them did. (The rewards for cheating were high: you were a "success".) Taylor soberly reported all kinds of fraudulent cutlery-bending as if his tests had been completely waterproof when in fact, to even an amateur conjurer, they were patently permeable.

Similarly, Edgar Mitchell (b1930) has a doctorate in science from the Massachusetts Institute of Technology and was the sixth man to walk on the Moon. He later founded the Institute of Noetic Sciences, based in California, which explores "powers of consciousness" and "phenomena that do not necessarily fit conventional scientific models", and gives public speeches endorsing some cheerfully fringe ideas, such as extraterrestrial-hypothesis ufology (see page 219): "A few insiders know the truth, and are studying the bodies that have been discovered."

Countless other examples could be cited; many will be found in the pages of this book.

It is of course the case that some elements – perhaps many – of the science of today may become the discarded science of tomorrow: the trick is knowing which elements those are, something virtually impossible to do from within one's own timeframe. The process of discarding ideas in favour of new and better ones is part of the healthy growth of science.

Spend 20 minutes surfing the World Wide Web and you'll realize that there are thousands of voices demanding that humanity take on board one new paradigm shift or another: if only we would accept that the answer to life, the Universe and everything is 42 then all the rest of human understanding would fall neatly into place. As Douglas Adams's parodic statement of the sum of all knowledge suggests, almost all of such claims are nonsense: many of them have been considered and rejected long ago, for good reason, and there is equally good reason why almost all of the rest have never been seriously considered. It is astonishingly rare for a paradigm shift to be triggered from outwith the scientific community, and it's not hard to see why: in almost all cases, no matter how much amateur theorists may batter against the wall of scientific indifference – like angry wasps against a window – the reason their theory is not being taken seriously is that it has fundamental flaws that are immediately obvious to anyone with even just a modicum of extra knowledge that the amateur does not possess. It's no real wonder that amateur theorists often feel themselves persecuted by the "lords of ivory-towered academia", or whatever – a regrettable situation to which there seems no easy solution: as noted above, scientists have limited amounts of time they can spend dissecting each and every new hypothesis that to them is quite patently nonsense. They don't have time even to *read* all of them. And there's the major deterrent to entering into any dialogue with an amateur theorist: nine times out of ten, whatever the scientist says, the theorist will not listen. (For a dramatic example of this obdurate deafness in action, spend a few minutes perusing the "Wacky Evolutionists" section of www.objectiveministries.org.)

It's therefore within the scientific community that it's most interesting to look around for straws in the wind that might indicate future paradigm shifts. As a recent example, there's the work announced in early 2006 by physicists George Chapline of the Lawrence Livermore National Laboratory and Robert Laughlin of Stanford University – with colleagues including Emil Mottola of the Los Alamos National Laboratory and Pawel Mazur of the University of South Carolina – on the hypothetical celestial objects they've called dark energy stars.

Although the concept of black holes has been part of the cosmological mainstream for some decades, and has roots that go back far further than that, there have always been some problems with it. For example, quantum mechanics states that information can never be lost from the Universe, yet information (ordered matter or energy) falling past the event horizon of a black hole would indeed be entirely lost – at least from our universe. Again, according to quantum mechanics time can never be "frozen", yet the prediction of the black-hole model is that energy (light) will be stretched out infinitely at the event horizon so that, to an outside observer, it will appear to freeze there forever.

Chapline and Laughlin were working on superconducting

crystals and the phenomenon called quantum critical phase transition when they discovered an unexpected result: the spin of the electrons seemed to show time slowing down. For some reason this reminded them of the hypothetical situation at a black hole's event horizon, and so with Mottola and Mazur they reanalysed the way a massive star should collapse, but insisting that in this model it should do so according to strict quantum mechanical principles. They found the end product of such a collapse would be not a black hole but a "quantum critical shell" containing an energy-rich vacuum – and, notably, no singularity. This vacuum would have a strong antigravitational effect, just like the "dark energy" currently being posited by cosmologists as the cause of the Universe's expansion. There would still be, as per a black hole, a powerful gravitational field drawing matter and energy in from outside, but within the shell there would be a repulsive force – which would be able to eject at least some raw matter and energy back out through the shell. Much of the ejecta would take the form of positrons and gamma rays . . . and, wait a minute, there's a hitherto-unexplained excess of positrons at the centre of the Galaxy, where it has been supposed a supermassive black hole resides. Could this object instead be a supermassive dark energy star? Likewise, the spectrum calculated by the team for the gamma-ray emissions is very similar to that of the enigmatic gamma-ray bursts astronomers have for some while been studying.

The hypothesis can be used to tie together a couple of other important cosmological puzzles. The enormous energy release of the Big Bang would be expected to create countless miniature dark energy stars (just as it would create countless mini black holes under the current theory), and these would have exactly the predicted properties of the hypothetical dark-energy particles that make up the "missing mass" of the Universe. Even more intriguingly, the team calculated the strength of the repulsive vacuum energy there would be inside a dark energy star the size of the Universe and discovered that it matches the deduced value for the dark energy that cosmologists have invoked to explain the Universe's expansion.

The new hypothesis also predicts that infalling matter will

cause dark energy stars to radiate in the infrared. This infrared radiation should be detectable by new instruments coming online in the near future, so within a decade or so it ought to be possible to put the hypothesis to the test by direct observation.

If Chapline and the others are proven to have been correct in their speculation, we will witness a paradigm shift in our understanding of the nature of the Universe – not least because, if the Universe is indeed an enormous dark energy star, then it becomes legitimate to wonder if there could be anything outside it . . . and if we could someday penetrate the quantum critical shell.

The dark-energy-star hypothesis is a perfect example of how a good scientific hypothesis should be. It has been born from experimentation and direct observation. It explains a great deal that was outwith the boundaries of the original experiment. The theorists themselves have been able to point to a way in which one of its predictions can be tested, as a result of which testing the hypothesis will either stand or fall – or, conceivably, stand but in a modified version. Further, the theorists are perfectly content to abide by the results of that testing: they show no signs that they will cling to their hypothesis if the results prove negative.

This is where this hypothesis stands in marked contrast to all but a tiny few of the "paradigm-shifting" theories put forward by amateurs.

Defunct science, pseudoscience, the ideological corruption of science, antiscience, conspiracy theories, hoaxing, fraud – you'll find instances of all of them here. You'll also find examples of just plain, straightforward, honest mistakes committed by people who've had the best of intentions but somehow simply got things wrong. Some of those mistakes, I ruefully have no doubt, will be mine: let me ask you in advance to forgive me for any of my own misapprehensions, biases or erroneous preconceptions that have crept in. Like Oscar Wilde's piano player, I'm doing my best.

Fanciful portrait of Claudius Ptolemaeus
from Sebastian Munster's *Cosmographia*, 1550.

CHAPTER 1

WORLDS IN UPHEAVAL

THE TERM "Ptolemaic cosmologies" is generally taken to describe all those schemes of the Universe in which the planets and fixed stars are thought to be embedded in rotating spheres at whose centre lies the Earth. The term refers to Claudius Ptolemaeus (Ptolemy; *c*90–168), whose version of the scheme seems to have been almost identical with that of Hipparchus (*c*190–*c*120BC), whose writings and observations Ptolemy almost certainly plagiarized; but ideas of the heavenly spheres date back at least as far as Pythagoras and the music of the spheres – although later Pythagoreans seem to have thought the Earth travelled around the Sun, and the Sun in turn around some "central fire" (the Pythagoreans were secretive, so it's often hard to establish what they did and did not think).

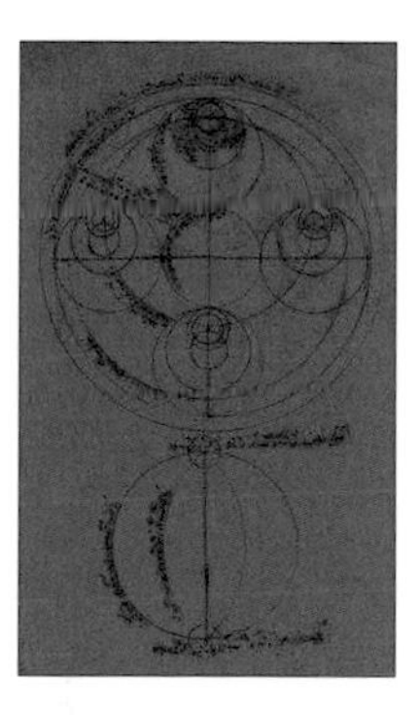

To Pythagoras (6th cent. BC), Plato (*c*428–*c*348BC), Aristotle (384–322BC) and many of their contemporaries, the Earth was a region of corruption and change while the heavens were a region of perfection and sublime stasis; phenomena which might disrupt this system – e.g., comets – were regarded as atmospheric events. But the perfect shapes were the circle and the sphere, and it would have been ridiculous to expect to find anything short of perfection in the heavens; thus it was axiomatic that the planets (including Moon and Sun) should trace circular paths about the stationary Earth; and, since it had been established long before by Anaximander (*c*611–546BC) that the fixed stars were embedded in the celestial sphere, it seemed obvious, too, that the planets were embedded in rotating spheres. (Anaximander also thought the Earth was cylindrical.)

It was manifest, however, that the planets did not straightforwardly travel in circular paths around the Earth. In particular, the planets Mars, Jupiter and Saturn displayed a distressing habit, at certain times, of stopping in their regular courses, travelling *backwards* for a while, and then continuing in their original direction once more – in other words, they looped the loop against the backdrop of stars. Various attempts were made to explain this while retaining the principle of uniform circular motion. Notable was that of Eudoxus (408–353BC), who considered that the poles of the sphere of the Moon were embedded in the sphere of Mercury, whose poles were embedded in the sphere of Venus, and so on. The system was elegant and ingenious, but did not in fact work very well.

It should be noted that not all the Greek philosophers agreed with the idea of spheres upon spheres upon spheres. Heracleides (*fl*500BC) observed that Venus and Mercury are never seen far from the Sun in the sky, and proposed that they might travel around the Sun – although it and all the other planets still circled the stationary Earth. But Aristarchus of Samos (*fl*270BC) produced a cosmology very similar to that of Copernicus, some 1800 years later. Needless to say, Aristarchus's ideas were largely ignored by his contemporaries.

The spheres were getting out of hand. Eudoxus had required 26 while a student of his, Callippus (*c*370–*c*300BC), realized that at least 34 were required. By the 2nd century BC the time was ripe for some sort of simplification, and this was supplied by Hipparchus. He proposed that there were indeed only the seven spheres of the planets contained within the sphere of the fixed stars. *But:* (a) the spheres' centre was not the Earth but a point in space which circled the Earth (the idea of the "eccentric"); (b) the planets were embedded in small rotating spheres (epicycles) whose centres lay on the larger spheres (deferents). By juggling the figures suitably, a reasonable approximation to the planetary movements could be achieved.

Ptolemy took Hipparchus's system, made it more complicated and even less plausible, and wrote about it in his *Megale Mathematike Syntaxis* ("Great Work of Mathematics"). The Arabs called it *Almagest* ("The Greatest"), and when this lost work was translated into Latin around 1175 it had an immediate impact upon European scientific thinking. Indeed, due to the Ptolemaic system's favour by the Church, it could be dangerous to criticize it. Nicholas of Cusa (1401–1464) got away with doing so, and was actually appointed a cardinal (his cosmology is startlingly modern); but Nicolas Copernicus (1473–1543) seems to have been too terrified to publish his theory until he was near death, and even then his printer introduced a preface to the effect that Copernicus's system was only an aid to better computation of planetary positions, not a description of physical reality. And, famously, Galileo (1564–1642) got into trouble over it.

For a long time historians thought that, because of the Church's insistence that the Earth was at the centre of God's creation, the reason the medieval cosmologists clung to the notion was for similar reasons of human self-aggrandisement. Some more recent historical researches, however, suggest the opposite was true: that many medieval cosmologists regarded our world, lying as it did beneath the heavens, as being, as it were, a sort of cosmic garbage dump; everything base fell as far as it could go, which was here.

The astonishing thing is that, while Copernicus's system was a closer approximation to the reality than that of Ptolemy, it was in fact poorer at predicting planetary positions than the old system of epicycles. This was one of the reasons why it took so long to be accepted. Another was that, Copernicus still being enslaved by the idea of uniform circular motion, he, too, had to make use of epicycles, just like Ptolemy – and he required a total of 48 of them, while the most sophisticated Ptolemaic system then current required only 40. This may have been one of the reasons why that superb observational astronomer Tycho

Brahe (1546–1601) refused to accept the Copernican system, devising instead his own modified version of the Ptolemaic cosmology.

Thus it was not until the work of Johannes Kepler (1571–1630), and his realization that the planets orbit the Sun in ellipses, not circles, that the Ptolemaic cosmologies slowly died.

THE MUSIC OF THE SPHERES

The Pythagoreans had two main preoccupations, music and number. The two interests are of course closely linked: if you halve the length of a vibrating string, the resulting note is an octave above the original; a string two-thirds the length of the original will produce the fourth of the first note you got; and so on.

In the initial, geocentric cosmology of the Pythagoreans, each of the planets (Moon and Sun included) was embedded in a sphere whose centre was the Earth; the outermost sphere was that of the "fixed stars". These spheres rotated at their different rates, and as they did so they whirred, each on a different note. Here, then, was for the Pythagoreans a marvellous mixture of music and mathematics – for the relative diameters of the spheres also played their part. According to Arthur Koestler (1905–1983) in *The Sleepwalkers* (1959):

> . . . the musical interval formed by earth and moon was that of a tone; moon to Mercury, a semi-tone; Mercury to Venus, a semi-tone; Venus to Sun, a minor third; Sun to Mars, a tone; Mars to Jupiter, a semi-tone; Jupiter to Saturn, a semi-tone; Saturn to the sphere of the fixed stars, a minor third.

(Uranus, Neptune, Pluto and the asteroidal bodies were of course unknown to the ancients.)

The idea of the music of the spheres has rarely been taken seriously since, although Koestler notes its frequent appearance in poetic writings. However, Johannes Kepler, in an early attempt to explain the workings of the (heliocentric) Solar

System, seems to have had a fundamental belief in the idea, marrying it to his theory that the orbits of the planets were related to the five regular convex polyhedra (solid shapes with flat faces which are all identical; e.g., the cube has six faces, each an identical square). Thus, according to Kepler's system, working outwards from the Sun we have: sphere of Mercury; then an octahedron (eight faces); sphere of Venus; then an icosahedron (twenty faces); sphere of Earth; then a dodecahedron (twelve faces); sphere of Mars; then a tetrahedron (four faces); sphere of Jupiter; then a cube (six faces); and finally the sphere of Saturn. However fanciful, Kepler's system has charm – and, as a way of estimating the relative diameters of the planets' orbits, is not especially inaccurate.

THE AGE OF THE EARTH

Do not believe *anything* you have thought you understood about our own home planet: somewhere along the line there has been a hypothesis that contradicts it. This goes right down to the most fundamental levels. To pluck just one example from the air, in 1947 a German amateur astronomer, Valentin Herz, was able to prove to his own satisfaction that the Earth does not rotate in the direction west-to-east, as we might have thought from the Sun's dogged habit of rising in the east. Instead it rotates east-to-west. Why this should be so has, alas, been lost to history along with Herz's notes on his hypothesis.

In 1809–12 William Hales (1778–1821) published *A New Analysis of Chronology*, in which he was able to list more than 120 attempts to establish the age of the Earth from Biblical studies carried out over the previous couple of centuries. The dates given for the Earth's creation varied between 6984BC and 3616BC. These estimates were produced largely as a result of counting the number of generations in the Bible and arriving at a figure of about 130. In order to establish the mean duration of a generation, and thus an age for the Earth (which was, remember, only a few days older than humankind), one had to call upon such disparate disciplines as mathematics and divine inspiration.

Why was it that even post-Renaissance thinkers could have believed the Earth so young? Essentially, because they lived in a society in which it was taken as an axiom that the Bible was literally true. But there was more to it than simply worship of God; involved, too, was the worship of Man, the highest of God's creatures. Surely God had required only a very short time to create the home for his masterpiece – indeed some were severely puzzled as to why it had taken him as long as six days to fit out the Universe for us.

This view of the youthful Earth was in stark contrast to those of the Classical thinkers, many of whom believed the Earth had always existed, and always would. Indeed, in some ways the ancient Greeks knew far more about the Earth than did their mediaeval counterparts. For example, Aristotle estimated the circumference of the Earth as about 75,200km (the modern value is about 40,000km); in the 3rd century BC Eratosthenes (*c*276–194) performed his famous experiment and arrived at a value of the order of 46,500km; and in the 1st century BC Poseidonius (*c*135–*c*51BC) produced an estimate close to 44,000km. While these estimates were not accurate, the point is that they were made, and were not ludicrous: it is ironic that the ancients not only realized the Earth was spherical but made quite reasonable measurements of its circumference, while as much as 1500 years later much of Western opinion was that the Earth was flat.

The most famous of all the chronologies is that of Archbishop James Ussher (1581–1656), produced in 1650–54; it is still sometimes to be found in annotated editions of the Bible. According to Ussher, the Earth was created in 4004BC. One powerful argument in favour of such a young Earth – in contrast to all the arguments against it – was that mankind was still expanding into new territories. Surely, if the Earth were much older than about 6000 years, all the available territories would by now have been occupied? Moreover, inventions such as printing were comparatively recent – why had they not been invented long before, if the Earth were indeed ancient?

In the 17th century there was, coupled with the idea of the

youthful Earth, a pessimistic view of what had happened since the Creation. It was the general feeling that the world had started off as a perfect place (God would not have created it otherwise), but, ever since, had been going to the dogs, and would continue to do so until the Second Coming. The deterioration of modern humanity in comparison with the ancients was regarded as self-evident: after all, went one line of reasoning, the ancients hadn't had to use spectacles!

This dismal theory held sway at the same time that the Earth sciences were poising themselves to burst into being, and so we find odd combinations of both notions – the old, theological, guilt-ridden past and the new, positivist, covertly agnostic, intellectual present. In *The Earth in Decay* (1968), Gordon Davies describes the Bishop of Gloucester drawing attention in the 1630s to the "debris-laden waters of a river in flood" – this was "indisputable proof of the reality of Nature's decay and of the mouldering of the very continents themselves". The Bishop was educated enough to recognize the process of erosion, but it merely reinforced his view that the world was in a state of degradation.

The young-Earth notion was long-lived. Even in the late 18th century the Comte de Buffon (1707–1788) was officially dressed-down for his suggestion that the Earth might be as much as 74,832 years old. In his private manuscripts he was daringly proposing that it might be as much as three *million* years old; wisely, he kept such heretical speculations to himself. Alarmingly, there are still, even in the 21st century, many who give credence to the notion of a youthful Earth. Most are to be found among the ranks of the Fundamentalist Christian movement, insistent as they are on the notion of a Divine Creation: we shall be encountering the Christian Creationists much further in Chapter 3. In the meantime, though, it should be stressed that, even among the Creationists, the young-Earth enthusiasts are in a distinct minority. But it's a noisy and on occasion depressingly influential minority.

Overleaf: The Earth newly created, from Sebastian Munster's *Cosmographia*, 1550

THE EARTH – FLAT OR HOLLOW?

A long-prevalent theory was that Man was a microcosmic replica of the Universe; since the Universe was, generally, regarded as geocentric, the "macrocosm" was, essentially, the Earth. Thus the eyes were the equivalent of the two great lights in the sky, the Sun and the Moon (eclipses were problematic), bones of rocks, hair of vegetation, pulse of tides, flesh of soil and earth, warts and boils of mountains, and veins of rivers. As metaphor, the idea has appeal; taken literally – as it was – it was of course extremely misleading.

For much of human history – although the idea has perhaps been far less prevalent than we might believe – the common notion has been that the Earth is flat. At least until recently there existed in the UK the Flat Earth Society, founded (*c*1900) in succession to the similar Zetetic Society. One theory which they had was that the Moon is only about 50km across.

The Babylonians saw the Earth as a disc floating at the centre of a sea around whose rim stood great mountains; upon these mountains rested the dome of heaven. In turn, the dome was surrounded by yet more water; leaks occasioned rain. The various celestial bodies entered and left the dome *via* convenient doors. The Egyptian view of the flat Earth was an elaborated version of this. The Earth was rectangular. The sky was the goddess Nut's star-spangled body, arched uncomfortably over the Earth; sometimes Nut took the form of a cow. Each day Nut gave birth to the Sun (which traversed her body by boat) and the stars. Later on, the Egyptians dismissed such fanciful notions of goddess-as-sky in favour of a vaulted metal lid around the inside of which ran a shelf. A river flowed along the shelf, and on it sailed the Sun and Moon gods; the stars were lamps hanging from the ceiling.

That these ideas were accepted is astonishing in light of the fact that as you travel north or south you see different stars come out at night – surely navigators such as the Babylonians *must* have noticed this effect. Still, although the ancient Greeks, as we have seen, were perfectly well aware that our world was a sphere, the flat-Earth idea survived. Martin Luther

(1483–1546) and St Augustine (d604) both insisted the Earth had to be flat, since otherwise people living on the underside wouldn't be able to witness Christ's descent on Judgement Day.

In 1895 the evangelical Christian and faith-healer John Alexander Dowie (1847–1907) founded the Christian Apostolic Church in Zion and the community of Zion, Illinois. About a decade later Dowie's leadership was usurped by Wilbur Glenn Voliva (1870–1942), under whose rule the community prospered – although its laws were strict: such offences as whistling on a Sunday could earn heavy fines. It was Voliva's contention that the Earth is flat: he travelled several times around the world trying to persuade other people to accept this. In his cosmology the north pole lay at the centre of a disc whose circumference was the southern "pole". Beyond the rim was Hades, but luckily there was a wall of ice (i.e., Antarctica) to stop mariners sailing over the edge. Voliva's flat Earth was motionless in space, with the Sun (small and nearby) and the stars revolving around it. The proof of the Earth's motionlessness was easy to come by: if the Earth were moving at great speed we should all be bowled over by colossal winds, as the atmosphere was "left behind". Voliva's flat-Earth system has probably been the most influential since the Middle Ages, representing little progress over the Babylonians.

The flat Earth lives on for a few – a very few – Biblical Fundamentalists, who can point to *Isaiah* ("He will bring back the scattered people of Judah from the four corners of the Earth") and *Revelation* ("Next I saw four angels, standing at the four corners of the Earth . . .").* Nevertheless, Fundamentalist US broadcaster Ian Taylor claims the whole notion of Christians perpetuating and even enforcing the dogma of the flat Earth throughout the Dark Ages is a myth – and that even the notion of the "Dark" Ages is slanderous, for those were really Christian ages during which, while ignorance, war and deprivation might indeed have held sway, there was tremen-

* Chris Morgan and David Langford pointed out in *Facts and Fallacies* (1981) that these statements do not necessarily imply that the Earth is flat and rectangular: it could equally well be a tetrahedron.

dous spiritual contentment because of Man's closeness to God. According to Taylor, virtually none of the early or Dark Age Christians believed the Earth was flat, only a few who rejected the notion of a spherical planet as part and parcel of their wholesale rejection of the ideas of the Classical philosophers. He points to Lactantius (245–325) and Cosmas Indicopleustes (6th cent.) as the sole examples of significant early Christians who maintained the world's geography must be as depicted in the Bible; the former was regarded at the time as a heretic by the Church Fathers, Taylor tells us, and thus had no influence on their thinking, while the latter's views were treated as at best fringe. The attribution by historians of a belief in the flat Earth to several centuries' worth of Christian thought is purely because Cosmas's opinions, once translated, have been taken as typical of the whole.

Taylor is guilty of a few omissions here. Christian thinkers like Pope Gregory the Great (c540–604) and St Isidore of Seville (c560–636), not to mention Luther and St Augustine (as noted), all maintained vehemently that the world was flat, with Gregory being in a position to enforce that view, which he did. As late as 1493 Pope Alexander VI (1431–1503) demonstrated his and the Church's belief in the flat Earth by proclaiming that the New World should be divided between Portugal and Spain, with the Portuguese having all the land to the east of a meridian and the Spaniards all the land to the west of it; it did not occur to the Pope that the Portuguese could sail eastwards and start claiming as much of the newly discovered continent's western coast as they wished. Prior to his circumnavigation of 1519, Ferdinand Magellan (c1480–1521) stated the current doctrine of the Church quite specifically: "The Church says the Earth is flat, but I have seen its shadow on the Moon, and I have more faith in the shadow than I do in the Church."

But the first great villain of all this, Taylor maintains, was Washington Irving (1783–1859), whose *The Life and Voyages of Christopher Columbus* (1828) created the picture of Columbus as almost alone in his belief that the world was round. (In reality, the sphericity of the Earth was well accepted by Columbus's time although, thanks to a miscalculation by Ptolemy, it was believed to be about one-third smaller than it actually is.) In 1834, just a few years after Irving's book, came an article by the French scholar Antoine-Jean Letronne (1787–1848) that depicted a Church which for centuries terrorized astronomers into ignoring the evidence of their own observations and adhering to the doctrinal view. This too, says Taylor, was a complete canard, although here he falls curiously silent about the particulars.

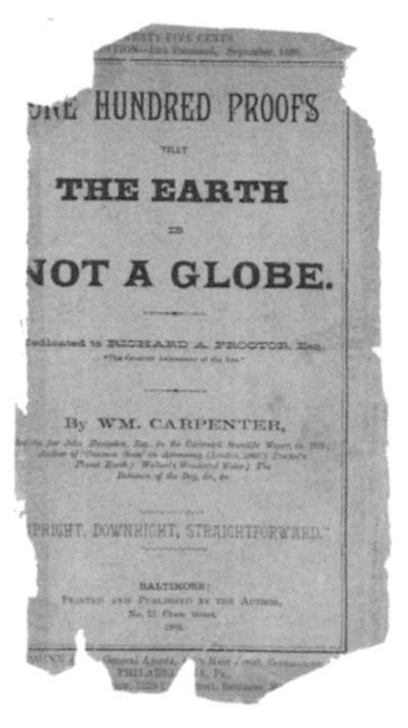
HUNDRED PROOFS

THAT

THE EARTH

IS

OT A GLOBE.

By WM. CARPENTER,

PRIGHT, DOWNRIGHT, STRAIGHTFORWARD."

BALTIMORE:

Further writers singled out by Taylor for especial condemnation include John Draper (1811–1882), author of *History of the Conflict Between Religion and Science* (1874), and Andrew White (1832–1918), founder in 1865 of the secular Cornell University and author of *History of the Warfare of Science with Theology in Christendom* (1897). Both men were essentially, according to Taylor's account of things, less scholars than political flagbearers whose interest was not so much in elucidating history as in advancing the cause of atheism. Taylor's attitude toward the motivations of the countless others who have come to tell much the same story, at least in the context of the Church's doctrine of the flat Earth, can be deduced from his overall categorization of them as "liberal historians".

Even though the ancient Greeks knew we live on a globe, they generally assumed the globe was solid. It was not to be until very much later that there appeared the idea of the interior of the Earth having a structure – core, mantle and crust.

A few centuries ago, however, various hypotheses began to emerge which had in common the notion that the Earth was not solid but hollow – or, at least, that there were plenty of hollow spaces within it.

The notion of the Earth's hollowness can be traced as far back as Plato and Aristotle and at least as far forward as

Edmond Halley (1656–1742), of Halley's Comet renown. The Earth is riddled with underground chambers and passages, through which waters flow or winds blow. Plato thought there might be a huge underground reservoir, which he called Tartarus. Sometimes the level in Tartarus rose, so that water was forced through the fissures and tunnels beneath the Earth's surface to debouch as rivers, lakes, seas, etc. Aristotle envisaged hot winds circulating through the fissures: occasionally these burst up through the surface as volcanic eruptions. Strabo (c60BC–AD20) pointed out that this was a good thing: without volcanoes to act as safety valves, the subterranean wind pressure might grow greater and greater until, one day, there would be a most terrible consequence.

In *The Mirrour of the World,* printed sometime after 1477 by William Caxton (c1422–c1491), it was not Aristotle's winds but the subterranean waters which were responsible for seismic effects. The waters flowed around turbulently, and often into closed underground caverns: the water pressure on the trapped air produced what we might term a popgun effect. This was the origin of earthquakes; volcanic eruptions required the generous admixture of quantities of Hell Fire and brimstone from below. (Later, in the 18th century, the Neptunists – see page 43 – would propose that volcanoes might be the outward manifestations of great subterranean coal fires.) In such terms, a strangely elegant model of the hydrologic cycle appeared. Water vanished from the seas through vast natural plugholes into the system of underground tunnels. As it swished around, heat from below (possibly from Hell) acted upon it; the net effect was that it was distilled, losing its saltiness. The distilled water flowed out onto the surface again at the tops of mountains.

That there was sufficient heat to operate this system was confirmed by the researches of the 16th-century Gabrielis Frascati Brixiani (c1520–c1582). He reasoned that the centre of the Earth was in fact the hottest place in the Universe, because, being at the centre of the Universe, all the heat rays from all the celestial bodies must eventually congregate there.

An odd yet durable theory was that sea-level must be higher than "land-level", because rock is heavier than water. Of course, "land-level" was observably being all the time lowered by erosion, which conjured up nightmares of, at some time in the future, no land remaining above the waterline. However, there were various explanations to offer comfort. Leonardo da Vinci (1452–1519) had the idea that the losses occasioned by erosion were adequately counterbalanced by the fact that eroded continents were lighter than before, and so floated higher in the water. That still left, of course, the alarming possibility of the landmasses one day disappearing entirely. Bernard Palissy (c1510–1590), in his *Discours Admirables* (1580), was one of several who noted this point. He said we should be very grateful that new rocks were always able to grow in place of the old ones.

The idea that the Earth might be entirely hollow has exercised a curious fascination over surprisingly large numbers of people during the last 150 years or so. In fact, there are two quite distinct theories involved.

The first is the conceptually simpler one. The Earth is spherical, but its interior is hollow. Inside is either a single great cavern or numerous layers of empty space – as if the Earth were a sort of spherical Russian doll. The theory's age is uncertain: in its modern form it seems first to have been proposed by US infantry officer John Cleves Symmes (1779–1829). Working on the assumption that the "access points" to the interior were at the poles, from about 1820 Symmes tried to mount an expedition to the north pole, getting so far as to put the proposition up before Congress in 1823. It seems that at first Symmes regarded the Earth as simply a hollow sphere, then later came to believe there were five concentric spheres, with ourselves living on the outermost.

His ideas were expanded in a novel, *Symzonia: A Voyage of Discovery* (1820), by "Captain Adam Seaborn" – this may have been Symmes himself. There had been earlier hollow-Earth novels – e.g., Ludvig Holberg's *Nicolas Klimius' Journey to the Underground World* (1741) and Casanova's *Icosaméron* (1788) –

and there were to be others, notably Edgar Rice Burroughs's *Pellucidar* series. Poe played with the idea, as of course did Jules Verne (*Journey to the Centre of the Earth*, 1864) and the authors of many tales of lost races. It is interesting to note that Symmes's son Americus realized that the "fact" of the hollow Earth explained exactly where the Lost Tribes of Israel had gone to: they went north and, on encountering the great polar hole, sort of fell in.

Symmes's theory has been revived from time to time. William Reed, for example, supported it in his *The Phantom of the Poles* (1906). Marshall B. Gardner, in his privately published *A Journey to the Earth's Interior* (1913; revised 1920), announced – four years *after* Peary's successful attempt to reach the north pole – that the aurorae are due to the internal sun shining through the holes onto our clouds. Raymond Bernard suggested in *The Hollow Earth* (1964), which leaned heavily on Gardner's work, that, not only did UFOs come from inside the Earth, they were piloted by Nazis, who had fled to this sanctuary in the last days of WWII. Brinsley le Poer Trench (1911–1995) was another to suggest a hollow-Earth origin for UFOs. In his *Secret of the Ages* (1976) he showed NASA satellite photographs of the polar regions. Some do appear, at very first sight, to indicate a great, angular hole at the north pole, but this is only because such pictures are made up of mosaics of photographs: the pictures leave black the areas that have not been photographed. The fact that pilots of high-flying aircraft on the polar route fail to notice a great hole was explicable, according to le Poer Trench, by the fact that compasses fail to function within about 250km of the poles: the pilots just *think* they're flying over the pole.

The only reasonably modern scientist of repute who seems to have considered the theory – although several natural philosophers came up with it in the 17th century – was the astronomer Edmond Halley. He thought the Earth might have a central core about the size of the planet Mercury, followed by two more shells about the size of Mars and Venus respectively, followed finally by the outer crust. The spaces between the shells were habitable, and filled with a luminous atmosphere.

Sometimes this atmosphere escaped at the poles, where the outer shell is thinnest – Halley knew about the polar flattening – and could be seen in the form of the aurorae.

The second main hollow-Earth theory is more complex. Could it be that we live on the *inside* of a hollow Earth? For this to be so, some assumptions would have to be made about the space in the interior of this curious shell. For a start, it has to contain the Sun, which *appears* to rise in the east and set in the west; and then there are all those distant galaxies . . .

This hypothesis seems to have been first proposed by the US mystic Cyrus Reed Teed (1839–1908; known also as Koresh). Teed proposed that the Sun had a light hemisphere and a dark hemisphere; we experience day and night depending upon which half of the rotating Sun is pointing towards us. To objections that at sunrise and sunset we can see the disc of the Sun, not just half a disc, Teed's response was that what we're seeing isn't the *real* Sun, just an "image" of it. Um. The paths of the planets required similarly impenetrable explanations.

Teed provided another of his "explanations" for the fact that, if you look directly upward at midnight, you do not see the lights of the cities on the other side of the world. He proposed that light could travel only so far around the concave

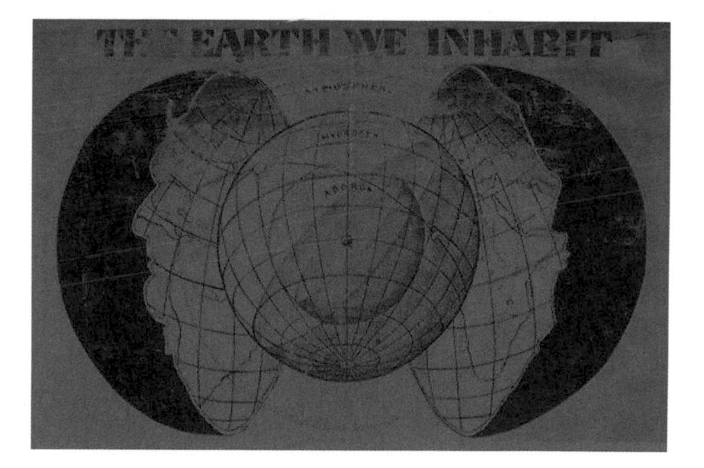

surface of the Earth before, as it were, taking a nosedive into the ground. Such a limitation did not apply to light of "arcane" frequencies – such as infrared. And so the Nazis, ever eager to find a new physics to which they could subscribe in order to discredit the "Jew science" of people like Einstein, conducted an experiment whereby they hoped to spy on the manoeuvres of the British fleet. They pointed their infrared telescopes up towards the sky at an angle of about 45°, and, to their surprise saw . . . clouds!

Teed's hypothesis did, however, have a more substantial effect on the development of Nazi technology. In 1933 the city council of Magdeburg decided it would test the hypothesis by asking the team of rocketry scientists then working in Germany – including Wernher von Braun (1912–1977) – to fire off a few rockets to see if these would land in the antipodes, on the far side of the hollow sphere. Unfortunately the sums forthcoming from the Magdeburg councillors proved to be not as great as expected and so the experiments had to be curtailed – by which time the best performance of the *Raketenflugplatz's* launches had been a horizontal flight of about 300 metres. Nevertheless, these experiments paved the way for the V1 and V2.

The idea of our living inside a hollow Earth did not die with Teed. In 1947, according to a *Time* magazine account in July of that year, the Argentinian amateur astronomer Antonio Duran Navarro derived it, or something very like it, seemingly independently. The Universe, Navarro announced, reportedly having spent years working on his ideas, was some 13,000km in diameter and contained entirely within the Earth.

Two further hollow-Earth hypotheses are worth noting.

Some very simple creatures are little more than a mouth at one end and an anal opening at the other, with a rudimentary digestive tract in between. It might be attractive to think of the hollow Earth in these terms, with its mouth and anus at the two poles – or so thought Alfred Lawson (1869–1954), founder of the knowledge system he modestly called Lawsonomy (see page 92). He held that the hole at the Earth's north pole acts as a sucking mouth, drawing in gaseous exhalations from the Sun and meteors, while the southern hole acts as the Earth's anal opening, blasting out the detritus. Maybe it's a mercy that, as le Poer Trench pointed out, aircraft do not in fact pass over the poles when their pilots think they're doing so.

The other relevant hypothesis is usually referred to as the Shaver Mystery. In the spring of 1945 the science-fiction magazine *Amazing Stories* published an article called "I Remember Lemuria", which seems to have been written by the magazine's editor, Raymond Palmer (1910–1977), based on an original by Richard Sharpe Shaver (1907–1975), a welder from Pennsylvania who had earlier published some science fiction. The readership response was considerable and, as further articles appeared, *Amazing*'s sales rose – to the extent that in June 1947 Palmer saw fit to devote an entire issue to Shaver articles. Some were collected in the book *I Remember Lemuria & The Return of Sathanas* (1948).

The main thrust of the "mystery" is that there exists within the Earth a race of detrimental robots, or "deros", who use ESP and secret rays to effect most if not all of the catastrophes to which the Earth is subjected. At one point the deros even took time off from crashing aircraft and starting wars to filch manuscripts from Palmer's desk!

All of this startling information was drawn from Shaver's racial memories. Later he was to recall from those same memories that the ancient astronauts who left the deros here in the first place, 20,000 years ago, are now returning. We see the vanguard craft in the form of UFOs.

THE EXPANDING EARTH

That the Earth might be expanding has sometimes been suggested in attempts to explain the fact that the continents were linked in past ages: if we imagine the Earth as a slowly inflating balloon whose surface has been encrusted with mud, we can see that, as the planet expands, so cracks appear in initially coherent landmasses and the smaller units move apart.

Of course, with the arrival of the theory of plate tectonics and its explanation of continental drift, the hypothesis proved unnecessary.

Alfred Wilks Drayson (1827–1901) in *The Earth We Inhabit* (1859) calculated that the rate of increase in the Earth's circumference was about 1cm per kilometre per year, and predicted that, among other effects, the truth of his theory would be established by telegraph wires beginning to snap.

Another feature of the geological past has brought the theory some attention. In bygone ages the oceans have on occasion occupied a far greater proportion of the Earth's surface than they do now. Could this retreat of the seas from the land be due to the Earth expanding, so that the world's waters had to spread themselves ever more thinly across the surface of the planet? Unfortunately, in the geological record we find that there have been also transgressions of the seas over the land – inexplicable in these terms unless one is prepared to countenance a rather more frightening hypothesis, that of the *oscillating* Earth! The retreats and transgressions are in fact primarily due to fluctuations in the extent of the Earth's glaciation, with differing proportions of the world's waters being locked up in the form of ice.

Various unorthodox geologists have questioned plate tectonics, pointing out that Antarctica is ringed by constructive plate margins (where new crustal material wells up from the Earth's interior). Since half of this new material must be heading inwards, towards the centre of the ring, the question arises: where is it all going to? An answer can of course be found in the expanding-Earth hypothesis. Some of these geologists proposed that about 100 million years ago the Earth's radius was only 80% of its current value. This means that, over the last 100 million years, the Earth's volume has approximately doubled. However, if this doubling in volume has occurred every 100 million years throughout geological time, about 2900 million years ago the Earth's radius must have been only some 10 kilometres – and when the Earth had just formed it would have been only 500m across. Its density then would have been one-third of that of neutron soup, the superdense material of which pulsars are composed.

An alternative to these unpalatable figures is offered if we say that the Earth's expansion began only recently, or has recently accelerated. But that fails to explain the good evidence we have that the continents were drifting relative to each other from a fairly early point in the Earth's history.

THE FORMING OF THE EARTH

To the Creationist it's easy: God created the Earth (and all the rest), end of story. Some cultures have just assumed the Earth has always been here, and may indeed have insisted upon this as a doctrinal matter. Most, however, have had their own stories of the Earth's origin, usually tied in to the sex life of the gods.

Mythology aside, it's one of the very obvious scientific questions to ask: Where did the Earth come from? The first serious attempt at an answer came from Immanuel Kant (1724–1804), who proposed that the Sun and the planets, Earth included, had condensed out of a spinning interstellar gas and dust cloud, or nebula. Soon afterwards, working independently, Pierre Simon de Laplace (1749–1827) presented a far more sophisticated version of this idea, now known as the nebular hypothesis: applying the known laws of physics and chemistry, Laplace envisaged a spinning gas cloud collapsing under its own gravity and heating up as it did so, to form the Sun, with the planets condensing out of the leftovers in an orderly fashion.

This implied a relatively cool origin for the planets, a point that increasingly troubled physicists as the 19th century progressed. The principal alternative offered was the planetesimal hypothesis. Rather than condensing directly to form the planets, the collapsing nebula instead formed countless small solid or semi-solid particles – planetesimals – spinning in orbits around the Sun. Because of gravitational perturbations, these crashed into each other frequently, generating heat. By chance, some of the accumulations of planetesimals grew faster than others; the bigger accumulations, exerting a stronger gravitational attraction than the smaller ones, drew in yet more planetesimals, until we were left with the eight then-known planets and their moons – plus all the myriad asteroids in the asteroid belt, which represented larger planetesimals that somehow had been unlucky enough never to coalesce into a planet.

A variation of this was the Chamberlin–Moulton hypothesis, put forward by T.C. Chamberlin (1843–1928) and F.R. Moulton (1872–1952), which was widely accepted in the US – less so elsewhere – during the early decades of the 20th century. (The UK astronomer Sir James Jeans [1877–1946] independently derived a very similar notion.) This allowed for the formation of the Sun through the collapse of a nebula, just as with the nebular hypothesis, but sought to explain the formation of the planets through postulating a close encounter at some later stage between the Sun and another star. The two stars, the hypothesis suggested, dragged long ropes of material out of each other which, after the stars had drifted apart again, fragmented into planetesimals that eventually formed the planets. This scenario of course implied, completely in contrast to the nebular hypothesis, that planets must be relatively rare in the Universe, for close encounters between stars are almost certainly extremely uncommon. (Although binary stars are frequent in the Universe, it's difficult to see how any planets they might form could last long without falling back into one or other of the parent stars.) For a while the Chamberlin–Moulton hypothesis was bruited to have replaced Laplace's nebular hypothesis, but it slowly fell from favour, along with other hypotheses that relied on close stellar encounters. It clung on longer in the US presumably for patriotic reasons.

It's now accepted that Laplace got things more or less right, but that the planetesimal hypothesis was not entirely wrong. A cloud of gas and dust received a nudge, perhaps from the shockwave of a supernova, and collapsed to form the Sun with, around it, an accretion disc where accumulations of matter built up, with collisions playing an increasingly important part. That process of collision has not entirely finished: the huge meteoritic impact that put paid to the dinosaurs happened relatively recently in the scheme of things, and there's no reason to believe another could not happen at any time. It was not long ago, after all, in 1994, that the planet Jupiter swallowed Comet Shoemaker–Levy.

THE SHAPING OF THE EARTH

In the 17th, 18th and 19th centuries, there seemed to be abundant fossil and geomorphological evidence that the Earth had once been widely covered in waters of great depth, probably at the time of the Biblical Flood. From this developed the theory of Neptunism.

The architect of the theory was Abraham Gottlob Werner (1749–1817), one of the most influential Earth scientists of the early 19th century. He proposed that the Earth had once been covered by a primaeval ocean, from which all the crustal rocks had precipitated over a comparatively short period. From this primordial sea precipitated first the crystalline rocks, such as granite – the "primitive" rocks, in Werner's terms; then came "transitional" (metamorphic) and "stratified" or "*Flötz*" (sedimentary) rocks; and then finally the "recent" (alluvial) and volcanic rocks. In spite of extensive evidence to the contrary, Werner believed that volcanic processes were of little consequence, certainly in the sphere of rock production. The main flaw in the theory was that there was no way in which it could explain where this vast ocean had gone to. Werner himself simply ignored the problem.

The death-blow to Neptunism came with the advent in the 1840s of the Glacial Theory (see below), but the real damage had been done before that. The great rival to Neptunism in the early 19th century was Plutonism, a theory generally associated with the name of James Hutton (1726-1797). Plutonism – often known as Vulcanism – won the day.

Popular from the latter part of the 18th century onwards,

Plutonism envisaged a cycle of erosion and uplift. Heat from below the Earth's surface thrusts compact ocean-floor sediments upwards (outwards) to form new landmasses, rather as warming the air in a balloon makes the balloon swell. Rivers wash across these landmasses, carrying sediments down to the sea. In due course, these sediments are pushed upwards (outwards) yet again – and the whole process repeats endlessly. The original rocks of the Earth were of igneous – i.e., volcanic – origin; rocks are, of course, still being produced through volcanic events,

The theory clearly had much to commend it – most notably its conception of a cyclical and continuing process rather than Neptunism's unique precipitation of all rocks from a primaeval sea, and in its realization of the importance of volcanism in rock formation. Modern planetary researches stress even more the importance of those same volcanic processes that Werner regarded as trivial in shaping the Earth's surface and producing its "primitive" rocks.

But Neptunism and Plutonism were not the only players in the game of trying to establish what were the processes that shaped the Earth. Arguably far more influential was the battle between Catastrophism and Uniformitarianism.

By the middle of the 18th century, the new science of geology was fairly well established; in particular, naturalists were applying the Scientific Method when constructing theories about the nature of the Earth: they were employing extensive fieldwork, rather than merely hypothesizing *in vacuo*, as had the ancients. To many it seemed clear that the geological structures and debris which they discovered were mute evidence of mighty catastrophes which had taken place in past ages; in short, they believed that the past had been very different from the present, and that abrupt changes and unique events had taken place during the shrouded millennia. Moreover, there hadn't been *much* past: it was generally believed that the Earth was only a few thousand years old – and, similarly, it was likely to end in only a few thousand years' time. Thus, not only had there been little time in the past for slow geological change, there was comparatively little time in the future for such processes to continue. But all around lay the evidence that major changes *had* occurred: this could only mean that such changes had taken place suddenly, catastrophically.

Georges Cuvier (1769–1832), the father of comparative anatomy, was unwilling to accept any idea of evolution; it was in large part thanks to his influence that the evolutionary hypotheses put forward by Lamarck (see page 133) were relegated to general obscurity, not to re-emerge until after Darwin published *On the Origin of Species* (1859). At the same time, Cuvier could not help but recognize from his own palaeontological researches that the faunas of the past were distinctly different from those of the present. He therefore accepted the concept of extinctions, speculating that species could die out very abruptly, to be replaced by God with new improved versions: the better-designed elephant had supplanted the mammoth, for example. He thus envisaged a plethora of individual divinely driven species creations. This fit in well with his ideas of Catastrophism: it was at times of great geological disruption, again determined by Divine Will, that the species of the past had been extinguished. One such catastrophe had been the Flood; not only had it drowned all the mammoths, it was after the Flood that humans had appeared on the scene. In this, of course, Cuvier was dissenting from a literal interpretation of the *Genesis* story, in which there are obviously plenty of humans around by the time of the Flood, but the match seems to have been near enough to mollify his religious beliefs.

So great was the general belief in Cuvier's Catastrophism that when the first remains of what we now call Neanderthal Man (*Homo sapiens neanderthalis*) were discovered, in 1856, the unknown prehistoric hominid species was called Deluge Man. The bones were discovered in a cave some 18m above the water level of the nearby river, and so the notion was that Deluge Man must have drowned while the Flood's waters were still subsiding.

Another French religious Catastrophist at the time was Jean André Deluc (1727–1817). He found evidence for the Flood in the rock strata, identifying a particular set of strata which he dated to 6000 years ago and called the *diluvium*.

Even so, ideas of a young Earth were beginning to look increasingly untenable in light of the evidence of the rocks and fossils. Religious scientists – and of course almost all scientists *were* religious at the time – came up with a new concept whereby the evidence in front of them could be made not to conflict with the testimony of the Scriptures. This was the idea of the world having had two ages, the ancient and the modern. During the ancient age, which might have been almost immeasurably long, God prepared the world for the arrival of the pinnacle of Creation, humankind; it was during this period that there occurred all the great catastrophes that Cuvier and others detected, and it was to this age that belonged all the bizarre extinct creatures evidenced by the fossils. The ancient age was thus the province of geologists and palaeontologists. The beginning of the modern age was signified, of course, by the appearance on the scene of humankind, and thus this age, deemed to have lasted only a few thousand years, was rightly the province of archaeologists. For several decades this division of interests fended off the inevitable clash between religious sensibilities and observed reality.

As early as 1797 John Frere (1740–1807) had discovered what he correctly identified as flint arrowheads that clearly were artefacts dating to long before humankind was supposed to have been present. His evidence was generally ignored as merely one of those puzzling little anomalies – besides, the stones were perhaps just naturally occurring stones that chanced to look as if they'd been worked – but the policy of "if we ignore it maybe it'll go away" couldn't last for long. The religious scientists tried, though. William Buckland (1784–1856), an important figure in geology's history despite his religious convictions, on being shown in Wales human bones mixed with the bones of a mammoth, concluded that the human, by definition a product of the modern age, must have been buried among the bones of a long-dead mammoth, a product of the ancient age, for superstitious ritual reasons.

Despite such inventiveness, the two-ages model was crumbling fast. Especially rickety, for both theological and scientific reasons, was the notion of the Flood as the demarcation between the two ages. When in 1859 the English archaeologists John Evans (1823–1908) and John Prestwich (1812–1896) compared notes with the French naturalist Jacques Boucher de Crèvecoeur de Perthes (1788–1868) about some tools he had found near Abbeville, it was immediately obvious that these were of the same ilk as those discovered back in 1797 by Frere. Boucher coined the term "antediluvian" – "before the Flood" – to describe the human culture responsible for the Abbeville axes. If humans could have existed before the Flood, perhaps they could have existed *long* before the Flood. This was still not inconsistent with *Genesis*, whose account said there had been people on Earth since very shortly after the beginning of Creation. But it was the evident enormous antiquity of Boucher's Acheulian culture (named later for a nearby church) that raised difficulties. Already many naturalists – Buckland among them – had started leaving the Flood out of their calculations altogether, even while still adhering to a two-ages model of prehistory. Now the stage was set for the admission that the Earth could not be a mere few thousand years old, and neither could humankind. By coincidence, it was also in the year 1859 that a book was published whose thesis required exactly such long tracts of time: Darwin's *On the Origin of Species*.

The Catastrophist view had much earlier received its first major challenge from James Hutton, in his *Theory of the Earth* (1788 and 1795), the bible of the new, rival faction of geologists, the Uniformitarians. Hutton saw the Earth as incredibly ancient, and maintained that throughout geological time the same slow forces had been at work upon it – erosion, denudation, gradual mountain-building. It was unscientific to postulate past catastrophic forces, said the Uniformitarians: "The present is the key to the past." Change was slow and change was gradual, but over the immensity of Earth history it could sculpt the mightiest mountainside – although, in overall terms, it changed the look of the world very little.

With occasional reverses, the contest went in favour of the

Uniformitarians. Charles Darwin (1809–1882) showed that the Catastrophists were right in thinking that life *developed* over time; but his theory implied slow change spread out over many millions of years. Baron Kelvin (1824–1907) threw the cat among the pigeons by "proving" that the Earth must be quite young – just a few million years old – but early in this century the discovery of radioactive materials let it be shown that the rate of the Earth's cooling must be many times slower than he had envisaged. But still . . . what had caused those eroded valleys in the mountains, if not the waters of the Flood?

In 1840 Louis Agassiz (1807–1873) had published his *Studies on Glaciers*, in which he showed that in past ages Europe had been far more extensively ice-covered than it now was. The concept of a past Ice Age was born. Here indeed was a catastrophe, but – and this is the crux – it was a slow, long-term catastrophe; moreover, it soon became evident that there had been not just a single period of glaciation, but several. The Ice Age (which we now call the Pleistocene ice age, having since learnt about earlier ones) was, therefore, part of the overall Uniformitarian pattern of agents-for-change at work upon the face of the Earth. And the new theory required vastly longer timespans than the Catastrophists were willing to allow.

Modern geologists and geomorphologists recognize that, in the great debate between Catastrophists and Uniformitarians, the irony is that both were right: had they been prepared to work out a synthesis of their ideas, rather than posture in righteous conflict, our modern view of the dynamic Earth might have emerged decades before it did. In the words of one modern geomorphologist (Brian John): "Of course, the major force shaping a mountainside is probably erosion, just as the Uniformitarians said. But a single landslide can do more to shape the mountainside in five minutes than erosion has done in thousands of years. And a landslide is a catastrophic event."

The Flood

Underpinning many of these outmoded geological ideas was of course the general acceptance that the Flood had been an actual historical event. That misconception still colours the ideas of modern Creationists, as we shall see in more detail in Chapter 3.

The Flood started on Sunday December 7 2349BC. The Noah family were, of course, well prepared: the Ark was fully stocked up with myriad animals; and fortunately it was the day after the Sabbath, so there was no need to miss church. Thus according to the cosy chronology of the Bible deduced by Archbishop James Ussher (1581–1656). Ussher added that on May 6 the following year (a Wednesday) the Ark came to rest on Mount Ararat.

The Old Testament version of the story of the Deluge is virtually identical with that in the *Epic of Gilgamesh*, except that there the inundation owes its origin to a decision by the gods to destroy mankind; the great god Ea is merciful, and warns Uta-Napishtim of the impending doom; thereafter Uta-Napishtim fills the Noah role.

It is a popular belief that, because the story of the Flood appears in mythologies all around the world, then "there must be something in it". A well known appearance is in Chinese legend. Here, the Flood was a primordial condition. As in Neptunism, the Earth was originally water-covered, and this caused various problems – agriculture, for example, was tricky. One man, Kun, was appointed to do something about it, and he planned to dam the waters using some soil stolen from Heaven. This theft was such a grave crime that he was promptly put to death. But some years later his corpse was cut open and out jumped a new hero, his son, Yu ("the Great"). Aided by various dragons, tortoises, etc., Yu was able to cut channels so that the water was able to drain off into the circum-terrestrial sea.

It is likely that Yu has the same historical status as King Arthur (although it is intriguing to learn that when he was born

The Flood imagined by Gustave Doré in 1865.

"his ears had three orifices"): a major figure locally at the time, transformed by legend into a major figure universally and for all times. Yu may have tackled the vast engineering project of building an embankment to prevent the Yellow River changing its course and flooding large areas of land. When tales of the Deluge reached the Chinese from the West, it was only to be expected that they should be used to augment an already exaggerated tale.

As late as the 17th century the Biblical account was taken by the great scientific thinkers to be literally true. In this, as in other instances (e.g., the Creation), discrepancies between the Bible's version and the results of scientific research were generally thought to stem from our imperfect interpretation of Nature, or of the Bible, or of both. The scientific acceptance of the Flood explained many observed geological phenomena neatly. Fossil shells and fish were found far above sea-level: the Flood. High in the mountains were found what seemed to be ancient river valleys: the Flood. The estimable UK geologist John Woodward (1665–1728) – who was brave enough to proclaim that fossils were organic remains when this view was not popular – judged the Flood responsible for the stratification (layering) of sedimentary rocks; he said it might have occurred as a result of a temporary suspension of the force of gravity. Clearly the Flood was important to his rivals, believers in Neptunism and Catastrophism, as well; indeed, the Flood was a stock tool of the Catastrophists well into the 19th century.

In the 20th century, the Flood largely disappeared from scientific thinking, but outside the scientific community there were plenty – aside from religious Fundamentalists – who gave its historicity credence. Hans Schindler Bellamy, a disciple of Hanns Hörbiger (1860–1931), gave a rather pleasing explanation for the Deluge. According to Hörbiger's World Ice Theory (see page 97), our Moon is but the most recent of several. Its predecessor spiralled in towards the Earth, impacting in prehistoric times. But before this crash, as it grew close to the Earth, it reared a huge "girdle tide" around the equator – a vast wall of water travelling at high speed; the moon was by this time circling the Earth six times a day, so the water must have been moving at about 8000kph.

Finally, of course, this moon disintegrated, showering the Earth with vast quantities of ice (Hörbiger's theory holds that all celestial bodies are thickly coated with ice) and rock. The melted ice from this moon and the waters from the collapsed girdle tide combined to flood the Earth. (One might think the girdle tide would already have so scoured the Earth that the subsequent invocation of a Deluge seems gratuitous.)

There are straightforward physical problems with the Flood story. If we assume the tradition of the Ark coming to rest on Mt Ararat is correct, then the waters must have covered the Earth to a depth of something over 5000m above sea-level. But, even if all the ice of both poles melted, sea-levels might rise by a mere 100m or so; add in all the other sources of ice, notably the freshwater glaciers found all over the world, and you might at a stretch get a figure of 300m. So the Flood enthusiast has the problem of explaining where all the water came from – not just that, but the equally great problem of explaining where it all *went to* afterwards. Some 19th-century adherents drew on Isaac Newton Vail's annular theory (see opposite) as a way of explaining where the water came from – the Flood legends were a record of the Earth's icy rings crashing down to the surface – but that still left the prickly matter of where it went to later. Into the hollow Earth, perhaps?

Then there's a zoological difficulty. Saving the animals "two by two" would not in fact have saved the species from extinction. Although the popular conception, formed largely by Hollywood, of a species' extinction is the death of one member of the final breeding pair, in real life species are doomed to inevitable extinction in the wild when their breeding population falls below a population that is typically in the high hundreds if not the thousands. Even if one argued that the Noah family could have constructed and maintained for decades a zoo for breeding purposes, it would have to have been one heck of a zoo.

Modern Creationists have the task of explaining the fact

that the fossils are found with the most primitive (oldest) fossils in the lowest rock strata all the way up to the seemingly most recent fossils in the youngest strata. It's hard to equate this with the belief that all the fossils formed at the same time and over a short period – i.e., during the Flood. To get around this snag, the Creationists typically introduce the concept of differential animal buoyancy. Those primitive-seeming creatures were the least buoyant, and so drowned first; higher forms of life were able to keep swimming longer. But this would mean that *all* of the marine lifeforms, like bottom-growing marine plants, should be found in the lowest stratum of all? Unfortunately . . .

And a final problem with the Flood account is that there is no evidence of such a global catastrophe in the *archaeological* record. Had there been a disaster that reduced the world's human population to a mere handful while destroying most cultural artefacts, it must inevitably have taken some while – centuries at the very least – to get civilization back up and running again, and the recrudescent civilizations would necessarily have differed considerably from the old, because no one ever rebuilds something exactly the same as it was before. Further, there would presumably have been only a single civilization, as opposed to the preceding plurality, because this time the Noah family would have had sole responsibility for getting things going. Yet the archaeological record shows the same human civilizations carrying on in a steady fashion, without any evidence of even a relatively minor disaster.

THE ANNULAR THEORY

In 1886 the Quaker scientist Isaac Newton Vail (1840–1912) put forward a theory which neatly tied together the Flood, the rings of Saturn, and the fact that the gas-giant planets Jupiter, Saturn, Uranus and Neptune have visible "surfaces" which look rather like incredibly thick "canopies" of clouds.

Once, Vail said, the Earth too had a ring or rings like those of Saturn – indeed, all planets pass at least once through an annular stage. He pointed out that the Earth's rings would naturally have lost momentum and collapsed towards the planet's surface: on hitting the atmosphere they dissolved into a tremendous canopy of water vapour whose lower level was some 150km above sea-level. The canopy became unstable near the Earth's poles; Vail thought the instability was due to there being less centrifugal force there, but later supporters thought it more likely the instability arose owing to interaction between the canopy and the Van Allen belts. Eventually the canopy, in its own turn, collapsed, first at the poles and then all over the planet. The result was the Flood.

Recent space probes have, according to Vail's disciples, provided dramatic confirmation. For example, the other gas-giant planets are now known to have rings, not just Saturn. And the surface of Mars shows dried-up water channels – obvious proof that the planet once passed through an annular stage. And there's the thick cloud canopy of Venus. While Venus is not a gas giant, it does have a tremendously clouded atmosphere: perhaps its rings collapsed not long ago.

But there are objections. First, to say that the gas giants have canopies shrouding their surfaces is risky: the gas giants don't really *have* surfaces. Second, the atmospheric layers we see are not made up of water gas: hydrogen, methane and ammonia are more the order of the day. Again, though there's some water in Venus's thick clouds, there's not very much – and the sulphuric acid, hydrochloric acid and hydrofluoric acid there are hard to explain in annular-theory terms.

The dense clouds in Venus's atmosphere have given rise to a runaway greenhouse effect on the planet. On the surface of Venus the temperature is around 500°C. If the Earth once had a similar cloud canopy, it too would have suffered from the greenhouse effect. Although the Earth is further from the Sun than is Venus, so that the temperatures would not approach the same 500° mark, still they would have been high enough to *cook* all lifeforms on the planet's surface while also boiling away the oceans.

An interesting modern variant of the theory is presented by Carl Baugh, of the Creation Evidences Museum in Glen

Rose, Texas. His main focus of interest is the nature of the Earth's environment in the days before the Flood. As a Creationist, he accepts that there were giant animals in the past and must find a means of explaining them in the very limited timescale of a recently created Earth. He posits that in antediluvian times the Earth was surrounded by a canopy consisting of thin ice layers sandwiching hydrogen with sufficient pressure that the hydrogen had taken on its metallic form. The canopy would have been held aloft by the magnetic repulsion between the metallic hydrogen and the Earth's core. In those days, he says, the Earth's magnetic field would consequently have been much stronger and also, since the canopy would act as a magnetic shield, far less liable to fluctuations caused by external sources. The stronger and more stable magnetic field could be expected to have, through influencing the electrochemical functioning of living cells, beneficial effects on our planet's antediluvian lifeforms: they could grow far bigger and live far longer – hence the exceptionally long lifespans of some of the patriarchs mentioned in the Old Testament, and hence also the size of some of the dinosaurs with which those patriarchs presumably coexisted. The canopy would also, as per a greenhouse, moderate and harmonize climates all over the world, which again would mitigate in favour of large, long-lived creatures and lush vegetation. A further effect of the canopy's presence was that atmospheric pressure back then was about twice what it is today.

Quite why *all* of the lifeforms of those times weren't huge and long-lived is unclear. And neither is it clear why, as with the Annular Theory proper, the greenhouse effect of the ice/hydrogen canopy wouldn't fairly soon elevate the planet's surface temperatures to oven-like extremes. Such considerations are unimportant to Baugh, whose main purpose is to devise some means of demonstrating the scientific accuracy of the Old Testament in the teeth of modern scientific knowledge.

Baugh has constructed gizmos – "hyperbaric biospheres" – to simulate on the small scale the conditions he believes pertained on the antediluvian Earth; there are claims that

NASA scientists are interested in these. Of course, should it prove that the combination of high atmospheric pressures and a strong, stable magnetic field does indeed have beneficial effects on living cells, this would be a matter of considerable interest; whether it would do anything to validate Baugh's hypothesis is less certain.

And there are other explanations for the longevity and giantism of ancient people and animals, one of them being the hypothesis known as Elective Polarity. The Earth's north pole at the moment points in the approximate direction of Polaris, a not especially bright variable star. However, owing to precession, the Earth's axis does not have a fixed orientation in space: like the handle of a spinning top, the north pole (and, of course, the south) traces out a circle, in a period of about 26,000 years. This means that some 13,000 years ago the northern pole star was Vega, one of the brightest stars in the night sky.

Vega would obviously be a much more majestic pole star than humble Polaris; and it was this which led to the suggestion by Frances Barbara Burton in *Elective Polarity, or The Universal Agent* (1845) that the great sizes attained by now-extinct animals were a result of the vigour of Vega being transmitted, somehow, to organisms here on Earth. We are, by comparison, puny little creatures because Polaris is such a puny little star.

POLE SHIFT

If the orientation of the Earth's axis should suddenly change, obviously direst catastrophe would be the inevitable result. Fortunately it's not likely to happen. However, various theorists have come up with pole-shift hypotheses in order to explain aspects of Earth's past. The basic idea is that the direction of the Earth's axis of rotation – which of course currently runs through the north and south poles – might all of a sudden alter so that you would find, say, Africa at the north pole and Antarctica on the equator. Pole Shift, insist the amateur theorists, is responsible for, among much else, the phenomena geologists call ice ages. Alfred Wilks Drayson (1827–1901), writing towards the end of the 19th century, was perhaps the first to point out that glacial remains in tropical regions were conclusive evidence of past Pole Shifts: the remains were clear proof that the Sahara once straddled the north pole. In recent years, recognition of the process of continental drift has rendered this aspect of the pole-shift theory rather superfluous.

In *Can You Speak Venusian?* (1976) Patrick Moore discusses at some length the pole-shift ideas of Dr Adam D. Barber, a US gyroscope engineer, founder of an organization called the Barber Scientific Foundation that seems to have done little except publish his one book. According to Barber's *The Coming Disaster Worse than the H-Bomb* (1954), the Earth has not one but two orbits: a short orbit of about 920 million million kilometres – an estimate which seems inaccurate by a factor of about one million – and a long orbit of about 16 million million million kilometres. In Moore's words: "He goes on to say that when the Earth's axis makes a right angle with both the large and the small orbits at the same time, the axis will shift suddenly by 135°, so that the North Pole will shift by 90°." The time taken for this sudden change will be of the order of 90 minutes.

That this sudden change is imminent is apparent. For one thing, it can be expected to occur every 9000 years or so, and some 9000 years have passed since its last recorded occurrence: the Flood. Moreover, look at a gyroscope. As the instrument "runs out of steam" its axis begins to precess – to trace out a larger and larger circle. Astronomers have known for millennia of the phenomenon of the Earth's axial precession, whereby our planet's axis does exactly this, taking about 26,000 years for each circuit. Clearly the Earth is about to fall over sideways! Barber suggested that it might be possible to save us from this holocaust by attaching suitable rockets to high mountain-tops on either side of the world.

It might come as something of a shock to find, in the mid-1950s, Fred Hoyle (1915–2001) and Thomas Gold (1920–2004) – two of the three originators of the Steady State hypothesis (see page 86) – putting forward superficially similar ideas. They suggested that the orientation of the body of the Earth with respect to the direction of its rotational axis is a result of both internal and external distributions of mass: if you suddenly stuck a great gob of matter on top of (say) London or New York, which are both well away from the equator, the Earth would slowly shift *on its axis* to bring this anomalously heavy concentration of matter closer to the equator. The Earth's equatorial bulge, which we have always considered to be a useful stabilizing device to avoid such disturbances, is, they said, not a permanent feature: the Earth can act in a fluid way, such that the bulge itself can move about to lie along each "new" equator.

Two distinctions are important between this idea and Barber's. First, Hoyle and Gold do not ascribe a period of as little as 90 minutes to the change; they suggest, instead, many thousands and possibly millions of years. Second, they stress that this change does not involve a shift in the orientation of the Earth's rotational axis; in Hoyle's usual elegant phrasing, a "turning of the Earth relative to its axis of rotation implies the same idea as the turning of the butter relative to the skewer [*Frontiers of Astronomy*, 1955]."

In 1980 appeared John White's book *Pole Shift*, which supported Barber's view in neo-Velikovskian style. The extent to which the pseudoscientists have misled the ordinary thinking reader can be assessed from the following extract from a review of the book by T.E.B. Clarke in *Punch*:

> That the ultimate disaster will be a shifting of the North and South Poles, causing the Earth to somersault in space, is a theory supported by soothsayers and scientists alike. It's happened before, they maintain, as proved by, for instance, the remains of mammoths and other prehistoric animals found preserved in ice in Alaska, the Arctic and Siberia, which show evidence of instant mass death while feeding on greenery where to our knowledge nothing green ever grew.

The misconceptions in this sentence are near-legion. Who *are* these "scientists" who are so nonmathematical that they are unable to calculate the amount of energy required to tip the Earth over? They're the people White describes in his book as scientists, that's who.

Just to confuse matters further, it's true that the Earth's *magnetic field* reverses its direction every few hundred thousand years or so. In other words, the magnetic poles swap: the "north" needle of your compass will one day point south. Such a change of polarity is long overdue; it's been an abnormally long interval since the last one. In the past, the polarity changes have often been marked by mass extinctions of species and the appearance, shortly afterwards, of new species; this is thought to be possibly because, during the changeover, more of the harmful radiation from the Sun is able to penetrate through to the Earth's surface. So there may indeed be a danger from "pole shift", although of the magnetic field rather than of the planet itself.

THE DRIFTING CONTINENTS

The idea of continental drift is today an accepted part of our understanding of the Earth. We know that the Earth's surface is made up of a number of fairly rigid plates, which are in motion relative to each other: their collisions and jostlings result in earthquakes, volcanism and mountain ranges. Only a few decades ago, though, such ideas were almost heretical. The 20th century had produced two important defenders of continental-drift ideas, Alfred Wegener (1880–1930) and Alexander Logie du Toit (1878–1948). Wegener was the pioneer; du Toit amassed a vast amount of evidence in Wegener's favour, and it does little credit to the scientific community that du Toit's consolidatory work was for so long either ignored or discounted.

Wegener's important book was *On the Origins of Continents and Oceans* (1915). Because he was not a geologist (he was a meteorologist), and because his theory by its very nature had to draw data from many disciplines, he made mistakes. Most of Wegener's errors were peripheral, but in the main the Earth scientists of the time did not realize this. They focused instead on the two errors he made that were major.

Why do the continents drift? Wegener sought to answer this perfectly reasonable question by invoking a new geological force, *Pohlflucht,* which encouraged continents to "flee" from the poles. In addition, he called upon tidal forces to explain such facts as that the Americas are moving westwards relative to Europe and Africa. However, there was no trace of the force called *Pohlflucht,* and tidal forces could not possibly be powerful enough to cause the effects Wegener desired. Since the mechanism was nonsensical, and some of the evidence at fault, it might be thought unsurprising that scientists rejected Wegener's hypothesis lock, stock and barrel . . . except that his primary evidence for drift was overwhelming. We've all noticed the "jigsaw fit" between South America and Africa; when you find that in addition there are rock successions in Africa which match rock successions in South America, and that, if the continents were put together, these successions would link like the teeth of two gearwheels, you naturally should wonder if perhaps Wegener had some right on his side. And then there was the palaeontological evidence. During the half-century or so between Wegener's proposal of continental drift and its acceptance by the majority of Earth scientists, some rather desperate attempts were made to explain why the fossil flora and fauna of South America, Africa, Australia and India showed great similarities. The most popular solution lay in hypothetical "land bridges".

At the beginning of the 19th century the Austrian geologist Edward Suess (1831–1914) had invoked a great but now sunken continent which had neatly filled all the holes in the Southern Hemisphere in the early days of the Earth, right up until about the middle of the Mesozoic (some 150 million years ago by today's reckoning); across this continent, which he called Gondwanaland, species would have been able to migrate between lands which are now separated by great tracts of empty ocean. Other geologists, balking at the concept of such a huge landmass having left so little surviving evidence, suggested instead a pattern of narrow isthmi (i.e., land bridges) linking up the various component parts of the jigsaw. Their case was helped by the existence of a contemporary land bridge, the Isthmus of Panama, and the good evidence that in comparatively recent times (a mere few thousand years ago) the Bering Strait had been cut off by a land bridge.

When the case for land bridges began to look dubious, but before the advent of the modern theory of plate tectonics (of which continental drift is an integral part), another suggestion was that guyots – submerged volcanic mountains, occurring in strings – were sunken island chains. Perhaps prehistoric faunas, including our own direct ancestors, had used the guyots as stepping-stones, taking the floras along with them? Alas, it is now known that guyots are produced not anywhere near coastal regions but right in the middles of the oceans, at mid-ocean ridges – whence they are carried, as on a conveyor belt, by the very process of seafloor spreading that drives the continents across the face of the globe.

The notion of a prehistoric network of land bridges has today been discredited: contrary to superficial appearances, the Isthmus of Panama is not an ancient survivor but has sprung up as the two American continents have travelled towards each other on collision course. The exception is the Bering land bridge, which is now widely accepted to have existed a few thousand years ago when the last glacial period of the current ice age caused worldwide sea-levels to fall. Across the isthmus thus created occurred migrations of all kinds of creatures, humankind included.

OUR NEIGHBOUR, THE MOON

To the ancients the Moon was just another planet, circling the Earth. However, they did realize it was the closest planet, and thus rather special. In recognition of this, many cultures worshipped the Moon, and most accorded it a fairly high position in matters occult. For example, according to Pliny the Elder (23–79), the belief that the witch-women of Thessaly could entice the Moon from the sky by ritual was so strong that the people would, at times of eclipse, set up an almighty racket in order to prevent the Moon's being able to hear the women's spells. The witch-women, of course, wanted the Moon down here so that its magical effects would be enhanced.

Pliny relays to us another charming hypothesis, this one concerning moonlight. Sunlight, as we all know, makes water evaporate, but by contrast moonlight "only causes water to evaporate with a rather gentle and imperfect force, and indeed increase its quantity".

Today we believe the Moon originated when, very early indeed in the Earth's history, while our planet was still forming, a body the size of Mars collided with it. (In the early Solar System such collisions cannot have been infrequent.) The ejecta from this enormous impact eventually became the Moon. An early form of this scenario was presented by George Darwin (1845–1912), the mathematician son of Charles. He proposed that, some 50 million years ago, the Moon was a part of the Earth. Then, for some reason, it split away; Darwin and his colleague Osmond Fisher (1817–1914) thought the Moon might have been pulled from the Earth by the gravitational attraction of a passing star, rather in the manner that the Chamberlin–Moulton hypothesis (see page 42) later sought to explain the origin of the planets as matter pulled out of the Sun by a close stellar encounter. Alternatively, according to Darwin and Fisher, it could simply have been centrifugal force that threw the Moon off. Although the Earth was still, according to their contemporaries, a largely molten ball at that distant time, the scar left by the Moon's departure could yet be seen, in the form of the Pacific basin. (In 1882 Fisher linked this theory with continental-drift ideas: the continents had been sundered at the time of the Moon's departure, and were currently drifting around in an attempt to readjust to the Earth's new shape.) When it was discovered that the Earth is very much older than had hitherto been thought, Darwin was – unwisely, with hindsight! – among the first to discard his own

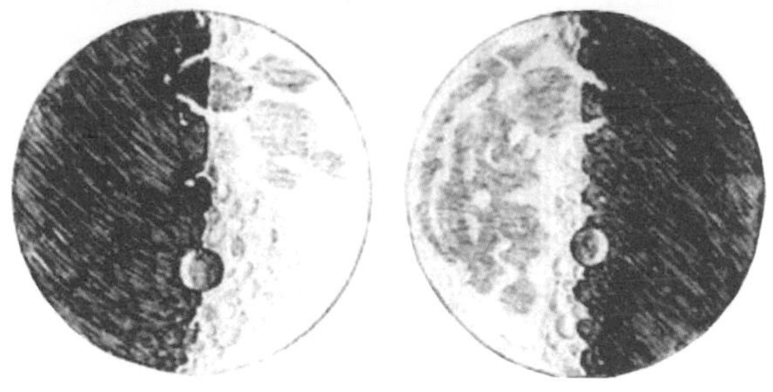

Upper: Engraving of the face of the Moon, done by Claude Mellan c1650.

Lower: One of Galileo's sketches of the Moon, from *Siderius Nuncius*, 1610.

hypothesis, although he still maintained that the Moon could be no more than about 60 million years old.

In 1919 the US astronomer W.H. Pickering (1858–1938) attempted to provide an explanation for the discrepancy between this age for the Moon and the contemporary geological estimate for the age of the Earth's crust of about 1200 million years. Clearly this was important because, while the idea of the Moon emerging from the molten surface of the Earth was staggering enough, the thought of it erupting from the *solid* surface was something more. Pickering's solution was elegant. He suggested that the solid material of which the Moon is formed did indeed fragment from the Earth over 1200 million years ago, but for the next 1140 million years or so it surrounded the Earth as a cloud of fragments – much as Saturn's rings surround that planet – before coalescing to form the Moon. Since the samples brought back from the Moon show that its surface is at least as old as that of the Earth, Pickering's idea has become irrelevant.

The genesis of the Moon's craters has been the subject of great debate since first they were seen by Galileo. There were two main schools of thought: one held that the craters were volcanoes, while the other maintained they had been caused by the impacts of giant meteorites during the early days of the Solar System. (We now know they are of impact origin.) But there were some theorists who belonged to neither school. Johannes Kepler put forward the idea that the craters might have been built by Moon-people. Some supporters of George Darwin thought they might be frozen "suck marks" (the sort you produce when you stand in tacky mud, then lift your foot abruptly) left over from when the Moon split away from the Earth. W.R. Drake likewise suggested that they might be nothing to do with volcanoes or meteorites: they were mute testimony to the evils of nuclear war – a war which in ancient times devastated the surface of that idyllic world.

The craters are not the only puzzling features of the lunar surface. There are also the dark patches, or *maria*. Harold Urey (1893–1981) – recipient of the 1934 Nobel Prize for Chemistry – suggested that the *maria* are the remains of organic material carried there as a result of the spectacular splashes produced as giant meteorites crashed into the oceans of the primitive Earth.

In the 19th and 20th centuries many theorists proposed that the Moon is a fairly small body, comparatively close to us. George Bernard Shaw (1856–1950) believed it was only 60km away – which would imply a lunar diameter of about 600 metres. W.R. Drake suggested that the Moon might appear to us much larger than it really is because our atmosphere might act like a giant lens, magnifying the image of the Moon by as much as twenty times. This would imply a diameter of perhaps 160km. By the time Drake proposed this idea, in 1964, several astronauts had already travelled in space, and none of them had reported that the Moon seemed smaller from out there.

Charles Hoy Fort (1874–1932), in his determined struggle against orthodoxy, decided to apply simple mathematics to the problem. He noticed that a large volcano on the Earth may be as much as 5km across. Assuming the Moon's craters were volcanic (most people did at the time), and ignoring any considerations of gravity, Fort guessed that lunar volcanic craters must be about the same size as those on Earth. Reassessing the face of the Moon in these terms, he deduced that it could be only about 160km in diameter – and thus about 18,500km away from us.

None of these theorists seems to have thought of the tides. If the diameter of the Moon is reduced to a twentieth, the volume is reduced by a factor of some 8000. Once you do the appropriate maths, you find that a Moon only 160km across and 18,500km distant would have to be about 20 times as dense as the Earth – in other words, 10 times as dense as solid lead! – in order to be able to produce the observed tides.

The Smallest Moon There Is prize goes, however, to a certain B. Bulstrode, who told us that the Moon is not really there at all. In his *The Moon is the Image of the Earth, and is Not a Solid Body* (1856–8) we discover that the light from the Earth is focused in such a way as to produce an image that we just *think* is the Moon.

THE PRESSURE OF MOONLIGHT

We've already noticed how some of the theorists who propose that the Moon is nearby and small forget the evidence of the tides. The picture is further complicated by other thinkers who refuse to accept that there is any connection between Moon and tides. For example, C.E. Last, in *Man in the Universe* (1954), informs us that, at the surface of the Earth, the Earth's pull is 288,000 times stronger than that of the Moon. Therefore, for the Moon to make any impression at all on the waters of the Earth, it would have to be no more than 65,000km away. "As regards the tide on the other side of the Earth, away from the Moon," he adds forbiddingly, "the position is still more impossible."

Various visiting ufonauts have confirmed Last's main theory. They have told lucky contactees that what actually happens is this: the weight of moonlight falling on the oceans pushes their surface down, so causing the water at the edges to spread out across the land.

On the subject of tides, it is of course because of Earth's tidal effect upon the Moon that the same face of the Moon (more or less) always points towards us. As the Moon goes around the Earth, then, it has to turn on its axis once a month. The 19th-century UK mathematician Henry Perigal (1801–1898) refused to believe this, and apparently spent much of his life attempting to persuade other people to agree with him. Rather touchingly, his obituary in the *Monthly Notices of the Royal Astronomical Society* tells us that, to help him in his crusade, he "made diagrams", "constructed models" and even "wrote poems".

THE SECRET NATURE OF THE SUN

In 1798 Charles Palmer published in *A Treatise on the Sublime Science of Heliography* his proof that the Sun is made of ice. He had discovered that he could singe tobacco using a lens made of ice. The conclusion was obvious: the Sun, too, is a lens of ice;

it seems hot merely because it focuses the brilliance of God. Scientific history might have been altogether different had Palmer used a *glass* lens.

Palmer's hypothesis was merely the most extreme of a cluster that maintained the Sun was temperate, and possibly inhabited. A short time later, Edinburgh Lawyer John Finlayson was to show that the stars, too, are made of ice and, for some unelucidated reason, are oval. In *Can You Speak Venusian?* (1976) Patrick Moore introduced us to several 20th-century temperate-Sun theorists, most important of whom was the Reverend P.H. Francis, who curiously was not scientifically unqualified: he had an MA in mathematics from Cambridge University. He used the concept of the temperate Sun to construct a whole new cosmology. (The following account is based on that in Moore's book, with his kind permission.)

The Sun is not on fire: if it were, its shape would constantly be changing, but we see a smooth disc. To think that our Earth is warmed by the Sun's heat is naïve: if heat cannot traverse the thin wall of vacuum in a vacuum flask, how could it pass through 150 million kilometres of vacuum to reach here from the Sun? The Earth, then, is warmed because the Sun (probably *via* its electromagnetic field) excites the atoms in our atmosphere into electrical activity, and this heats things up. Anyone can test this by climbing a mountain to where there are fewer atmospheric atoms to be excited, and noticing that it's colder there.

The cosmology which Francis has constructed is both fascinating and elegant. Since the Universe is infinite, the Earth must be at its exact centre – just like everywhere else must be at the Universe's exact centre. (This is the converse of the relativistic idea that, for almost identical reasons, there is no such thing as a "centre of the Universe".) However, although the edge of the Universe lies at infinity, this does not mean that all points on the boundary are an equal distance from us: while all are an infinite distance away, some are at a less infinite distance than others. Thus the Universe can be envisaged as surrounded by crinkled aluminium foil, to use a homely metaphor. The cool Sun is the only actual star; but its light is reflected back from the foil as countless sparkling points of different brightnesses, which we think are all the other stars. Sometimes the Sun's light penetrates the boundary of the Universe, swoops around, and then reenters the Universe to look like an extremely faint star.

The relationship between Francis's cosmology and his concept of the temperate Sun is not 100% clear, but it seems the coolth of the Sun was vital to the workings of the whole system.

Sir William Herschel (1738–1822) was one of the most distinguished astronomers of all time, and also one of the more extreme believers in the plurality of inhabited worlds (see page 197): he considered that life on the Moon was an absolute certainty and life on all the other planets of the Solar System almost as assured. This was not at all an unreasonable stance in his age, when we had very little notion of the nature of the other planets, which for all people knew could be just like Earth. But that hardly seems to justify his odder pronouncement that, while the *surface* of the Sun is undoubtedly hot, there was probably beneath it a temperate land where dwelt intelligent beings. Moreover, a little application of logic revealed to him that the social structures of those beings were almost certainly more or less identical with our own. While it may seem surprising that such a giant of the history of science should believe this, what is truly astonishing is that another such agreed with him: Dominique Arago (1786–1853). It was as late as 1952 a German theorist, Godfried Büren, proved to his own satisfaction that the Sun was hollow and that there was within it a cool globe. This inner Sun was covered with lush vegetation and consequently appeared dark to us when we could see it through the occasional temporary holes that opened in the Sun's fiery outer layer; this was Büren's explanation for sunspots, which in turn supplied strong evidence for his hypothesis.

Büren was so convinced of the validity of his proof that he offered a prize of DM25,000 (then a small fortune) to anyone who could counter it. Great was his chagrin when a major German astronomical society went ahead and did just that.

Naturally, their rebuttal of the theory was deemed by Büren to be arrant nonsense, and so he refused to pay up – until, that is, they took him to court.

THE SUN'S UNKNOWN PLANETARY RETINUE

The French astronomer Urbain Leverrier (1811–1877) determined during the 1840s that there was something odd about the orbit of Mercury. The point in a planet's orbit when it is closest to the Sun is called its perihelion; Leverrier found that the perihelion of Mercury advanced around the planet's orbit by about two-thirds of a degree each century, a very small precession but one that could not be explained in terms of Newton's Law of Universal Gravitation. In 1845 Leverrier proposed that the discrepancy was due to a perturbation of Mercury's orbit by the gravitational attraction of a hitherto unknown planet orbiting even closer to the Sun than Mercury; he dubbed this hypothetical planet Vulcan, and the search was on to catch a sight of it. According to Leverrier's calculations, it must lie in an orbit about 30 million kilometres from the Sun, orbit once every 34 days or so, and be about 1500km across. Needless to say, sightings of Vulcan soon followed – the first seems to have been by French physician and amateur astronomer Edmond Modeste Lescarbault in 1859.

Alternative explanations were put forward for the advance of Mercury's perihelion, including that there might be an inner asteroid belt; but the most intriguing proposition was made by the distinguished US astronomer Asaph Hall (1829–1907). Hall suggested that the famous inverse square law of gravitational attraction did not quite hold: while Newton and all who had followed him assumed that, if the distance of a planet from the Sun were to be doubled, the Sun's gravitational pull on that planet would be reduced to one-quarter ($2^2 = 4$, and the inverse of 4 is ¼), Hall proposed that the pull might in fact vary in inverse proportion to just a little more than the square of the distance (2^{2+x}, where x is very small).

In 1915 Einstein's General Theory of Relativity successfully accounted for the advance of Mercury's perihelion, and since then it has been assumed by orthodox science that Vulcan does not exist. However, some adherents of the extraterrestrial hypothesis of ufology (see page 86) still claim that not only does Vulcan exist but it is of planetary size and its inhabitants pilot UFOs. They would have to be very thick-skinned inhabitants, however: the surface temperature on Vulcan could be expected to be about 1650°C.

The idea that there might be a twin to Earth, Antichthon, following the same orbit as our planet but always on the far side of the Sun from us is much older, dating back at least as far as the 17th century and probably earlier than that. In more recent years certain proselytisers for ufology's extraterrestrial hypothesis have cited it, in tandem with the hypothetical Fifth Planet, as a possible source of alien visitors; they generally call the planet Clarion.

One problem with any theory concerning Antichthon is that it is impossible for a counter-Earth to remain permanently hidden from our view by the Sun. The counter-Earth's orbit would be perturbed by the gravitational pulls of the other planets of the Solar System, in just the same way as is the orbit of the Earth. However, these pulls could never cancel out, as it were, in such a way as to keep the two worlds forever out of each other's view.

VENUS

For a long while Venus was regarded as Earth's "sister planet" – about the same size but a bit hotter (because much closer to the Sun). Its features are permanently obscured from us by impenetrable layers of cloud, so a common deduction was that it rained a lot on Venus. More recently it has been realized that surface temperatures on Venus are of the order of 500°C, which is hot enough that the surface rocks glow redly from their own heat; and that those clouds are made up not of water vapour and the like but of corrosive compounds such as sulphuric and hydrofluoric acids. In short, you could cook your dinner without the use of an oven, but your casserole would probably be dissolved the first time it rained.

An interesting feature of Venus is that, when the planet is seen as a crescent, the dark area is very faintly illuminated. (This can be seen also with the crescent Moon: in the Moon's case it is due to reflected Earthshine.) Most might think the illumination to be simply a result of refraction in Venus's atmosphere, but the 19th-century astronomer Franz von Paula Gruithuisen claimed it was a result of the Venusian custom of setting the forests on fire whenever a new emperor succeeded to the throne.

Emanuel Swedenborg (1688–1772) was another unorthodox theorist interested in Venus. Thanks to psychic communications, he discovered there were *two* intelligent races on Venus, each occupying a hemisphere. One was gigantic, cruel, savage, avaricious and gluttonous, the other deeply spiritual.

Because early estimates of Venus's surface temperature varied so widely, countless theorists speculated quite sensibly about life on Venus. The high surface temperature and the lack of, for example, oxygen in the atmosphere might now seem to militate against there being Venusians – but in 1959 V.A. Firsoff asked if it were really true that Venus's atmosphere had very little oxygen: might it not be that the planet's magnetic field retained oxygen at low levels in the atmosphere, below the obscuring clouds? A reasonable suggestion, until almost immediately *Mariner 2* showed that Venus has no magnetic field to speak of.

That Venus might have been born as a comet spat out from the surface of Jupiter is a central tenet of Velikovskianism (see page 78).

MARS

The main excitement about the planet Mars is that, obviously, it is the home of a civilized race. The birth of this idea was part of the suddenly popular belief that perhaps *all* celestial objects were inhabited – the Moon, the Sun and the stars included.

The real genesis of the civilized Martians came with observations made by the Italian astronomer Giovanni Schiaparelli (1835–1910) in 1877–81. He declared he could see straight markings criss-crossing the Martian disc, and dubbed these "*canali*", meaning "channels"; the word was rapidly mistranslated into English as "canals", and the story began. Obviously these canals had been built by an ancient Martian race in an attempt to bring water from the polar icecaps to irrigate the drought-scourged deserts.

The two chief champions of the hypothesis were Camille Flammarion (1842–1925), who believed all worlds were inhabited and who had reported observing the growth of vegetation on the Moon; and particularly the US astronomer Percival

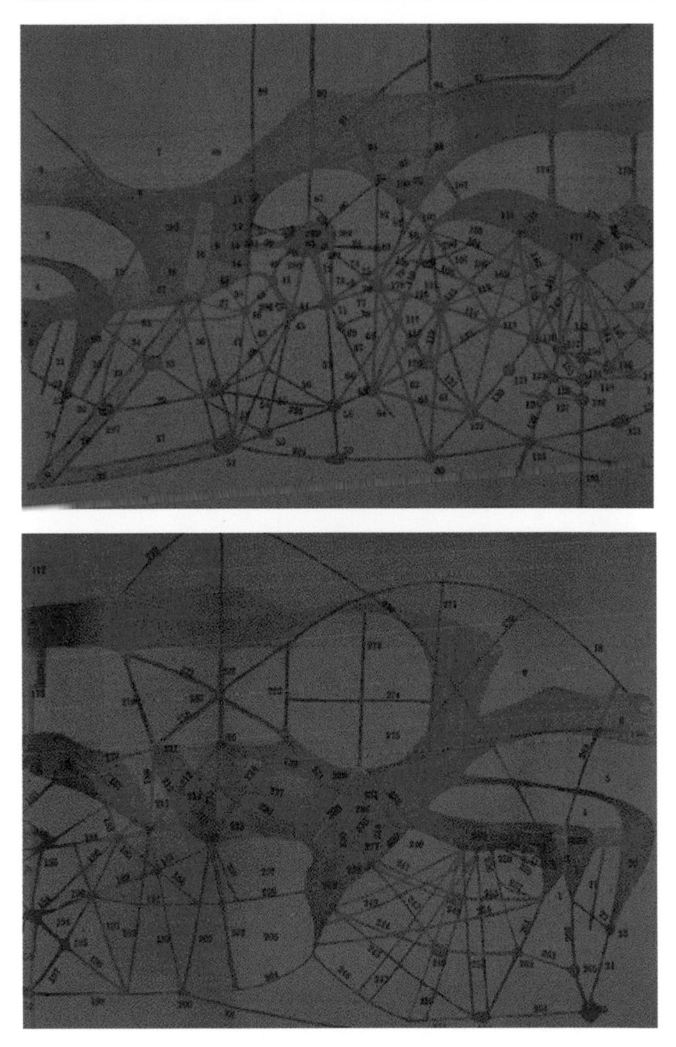

Facing page: The Martian canals depicted in Camille Flammarion's *La Pluralité des Mondes Habités*, 1862.

Above: Two of Percival Lowell's many maps of the Martian canals.

Lowell (1855–1916). For 15 years Lowell studied Mars, mapping the canals incessantly; his books *Mars and Its Canals* (1906) and *Mars as the Abode of Life* (1908) were enormously influential, not only among some astronomers but also, more significantly, among the general public. H.G. Wells's *The War of the Worlds* (1898), although of course purely a work of fiction, also played its part. Both Nikola Tesla (1856–1943) and Guglielmo Marconi (1874–1937) claimed to have received radio signals from Mars.

As early as the start of the 19th century – long before Schiaparelli and the rest – there were plans afoot to dig colossal geometrical figures in the Sahara in order to communicate with the Martians. Those Martians would have needed powerful telescopes: it was not until 1971 and the *Mariner* 9 probe that we identified a dot occasionally observed from Earth as the volcano Olympus Mons, which is at its base nearly 650km across.

Later in the 19th century the French inventor Charles Cros (1852–1888) picked up a suggestion from Carl Friedrich Gauss (1777–1855) that we could build huge fires on the Sahara, laid out such that they represented a geometrical theorem – e.g., the Pythagorean theorem. Cros attempted to get funding from the French Government for such a project. The French Government also declined to fund his scheme of building a very large burning-glass with which he could focus sunlight in such a way as to scorch out messages on the Martian desert.

The most famous Martian architectural feat of all, the notorious Face on Mars, is located in the region of the planet's surface known as Cydonia, about 40° north of the equator. It was first spotted by astronomer Tobias Owen in a photograph taken in 1976 by the *Viking* orbiter. The area in the photograph, which was shot from an altitude of about 1600km, is about 56km x 50km. A second *Viking* photograph, taken 35 Martian days later when the angle and lighting conditions were different, permitted further analysis to be done of the structure: it is about 2.5km long and about 2km wide, and stands about 600m tall above its surroundings.

About 15km away, in the same photograph, is what looks like a severely damaged five-sided pyramid, a structure called the D&M Pyramid after Vincent Di Petro and Gregory Molenaar, the two people who identified it. US Face enthusiast Richard Hoagland (b1945) claims to have identified numerous additional "artificial" features in that original *Viking* photograph, and suggests it must be more than coincidence that so many of them should be crowded into such a small area of the red planet's surface: surely their proximity alone must go a long way towards proving his hypothesis that all of these structures are vast monuments created by a sophisticated Martian civilization?

Other researchers have of course got in on the act, analysing other photos of the Cydonia regions, and identifications of further seemingly architectural objects have proceeded apace. And Cydonia is not the only region of the planet to have drawn attention: in Elysium, on the opposite side of the planet, photographs taken five years earlier, in 1971, by *Mariner 9* show what could be read as a group of three-sided structures that stand an impressive 1.6km high.

Cut to 1996 and the analysis of a meteorite found in Antarctica and shown to have originated on Mars during some major impact in the early days of the Solar System; after floating around in space for several billion years, this stone is thought to have finally plummeted to ground in Antarctica some 13,000 years ago. The investigating scientists found what appeared to be organic remains within the meteorite that would suggest life on at least the unicellular level might have existed on Mars some billions of years ago – at roughly the same time that, on Earth too, the highest level life had reached was unicellular. The results are still regarded as equivocal, since there is a possibility that the meteor was contaminated by terrestrial life during the 13,000 years since its arrival on Earth; indeed, most scientists believe this is probably the case . . . while admitting they could be wrong.*

* A further Martian-origin meteorite, apparently bearing organic traces within it for which contamination seems an implausible explanation, is being analysed at the time of writing.

Face enthusiasts are significantly less cautious in their estimation of the results. Life on Earth, they point out, has evolved very significantly over the past few billion years, and we should expect life on Mars to have done the same. Life here has produced us; is it not likely that life on Mars has produced creatures capable of building mighty monuments like the Face and the D&M Pyramid?

Yet there are precious few signs of that civilization still being extant today. What could have happened to them? Well, we recall that the D&M Pyramid looks as if it suffered extensive damage at some stage, so a Martian war, using Martian superweapons, seems a plausible explanation. More likely, however, is some sort of geological catastrophe. It is known that in the past the planet did indeed suffer at least one such, although exactly what it was, and exactly when, are imponderables. It is also regarded as fairly certain that in a past era Mars enjoyed a climate considerably warmer than it does today, with much more surface water present. Face enthusiasts – not unreasonably – put these two facts together and speculate that the catastrophe occurred during Mars's temperate period (or one of them), creating mighty flooding, volcanic eruptions and earthquakes sufficient to wipe out the ancient Martian civilization.

But perhaps not all of them. Could a few survivors have escaped the Sun's fourth planet and fled inward to its third? After all, are not the Great Pyramids of Egypt eerily reminiscent of the pyramids on Mars? Are there not legends that the Great Pyramids were erected by some primordial race as storage places for that race's important treasures against the ravages of the Flood? Is it not feasible that . . .?

Unfortunately for Face enthusiasts, most planetary scientists are fairly sure the "structures" on Mars are simply naturally occurring geological features, and that their apparent artificiality is merely a result of the "inkblot effect" (see page 202): if you look at a randomly formed collection of shapes with hopes and/or expectations in your mind as to what you will see there, then sure enough you're likely to see it. However, the Face legend lives on, and will doubtless continue to do so until – or perhaps even after – space probes carry out a closer investigation of the Cydonia region. Already the Mars Global Surveyor has, in 1998, taken far better photographs of the Face which make it appear significantly more like a natural feature than an artificial structure. This seems to have done nothing to dampen anyone's ardour.

At the very limits of visibility, the eye tends to join up what are in fact discrete images, and this is the most probable explanation of the canals: Schiaparelli, Lowell and all the other astronomers who saw the lines were merely the victims of an optical illusion. They were far from the first astronomers to be deceived in this way. Both Robert Hooke (1635–1703) and John Flamsteed (1646–1719) managed to observe stellar parallax, they claimed, long before instruments sensitive enough for any such measurement came into existence; having no conception of how far away were the stars, the two highly distinguished scientists claimed figures that were wrong by a factor of hundreds. But, even if one dismisses the canals as optical illusions, and the Martians as at best bacteria, the threat from Mars still cannot be discounted. Perhaps the most chilling possibility is the belief by certain Buddhist priests in Sri Lanka that the planet is responsible for venereal disease.

THE FIFTH PLANET

Further from the Sun, in the region between Mars and Jupiter, lies the asteroid belt – a collection of countless small bodies ranging in size from minute up to Ceres, about 750km across. Until quite recently it was seriously maintained they were the relics of a planet which had for some reason exploded or fallen apart at the seams. Today it is thought, on good evidence, that they are detritus which for some reason did not coalesce to form a planet – possibly a protoplanet was in the process of formation, but the gravitational pull of mighty Jupiter pulled it apart faster than it could put itself together.

Postwar advocates of the World Ice Theory (see page 97)

invoke not one but *two* planets between Mars and Jupiter. In *Can You Speak Venusian?* (1976) Patrick Moore reported:

> It is thought that in the remote past, a large planet, Phaeton, used to exist there. It was disrupted by a wandering body named Maldek or Mallona, which ran off course, bounced off Pluto and Neptune, knocked Jupiter violently enough to produce the gaping gash we now call the Red Spot, and finally smashed against Phaeton, breaking both itself and its victim into a shower of asteroids.

VELIKOVSKIANISM

The essential story of the catastrophic past told by Immanuel Velikovsky (1895–1979) seems to be – in places it is ambiguous and/or inconsistent – as follows:

About 4000 years ago, what is now the planet Venus was born as a comet, spat from a volcano on Jupiter. The new comet fell towards the Sun and in due course adopted a highly elliptical orbit which cut across that of the Earth. About 1500BC, or later, the Earth passed into the comet's tail. The first effect was that the atmosphere was filled with a fine red dust, which sank to the surface, staining the land and polluting the water. Worse was to follow. As the Earth came closer to the comet's head, larger particles fell, producing dramatic meteor showers – and even a hail of meteorites, which doubtless took its toll. Worse was to follow. Vast quantities of hydrocarbons rained into the atmosphere, some to reach the ground and percolate downwards to form our modern petroleum supplies; more terrifyingly, however, some of the falling petroleum was ignited by lightning, so that flame filled the skies. Worse was to follow. The Earth neared the comet's head, whose gravitational influence began to wreak havoc on the planet. The poles shifted (see page 56) and the crust was distorted. The oceans drained across the land, volcanoes erupted, hurricanes raged. The poles shifted again and again, and several times the Earth was locked motionless on its axis. Worse was to follow. Coruscating bolts of electrical energy flashed between the two bodies, flaying the Earth. Worse was to follow . . .

The Israelites, meantime, made use of the mayhem to facilitate their flight from slavery. The disruptions of the crust had raised part of the floor of the Sea of Passage (Velikovsky does not agree that this was the Red Sea – named, of course, because of pollution by the red dust mentioned above), and across the isthmus so formed the Israelites eagerly scurried. The armies of Egypt attempted to follow but, serendipitously, a colossal bolt of electricity from the comet's head struck the Earth, so that the isthmus was abruptly swallowed and the armies drowned, The Israelites followed the direction of the comet in the sky: it looked like a pillar of smoke during the daytime and a pillar of fire at night.

But in one respect the comet helped mankind: manna. According to Velikovsky, "The tails of comets are composed mainly of carbon and hydrogen gases." This is the reason for his assertion, above, that at one point petroleum rained from the skies. He proposes that, in the aftermath of the comet's passage, "Most probably the atmosphere [of Earth] discharged its compounds, presumably of carbon and hydrogen, in a similar way . . . After the nightly cooling, the carbohydrates precipitated and fell with the morning dew. The grains dissolved in the heat and evaporated; but in a closed vessel the substance could be preserved for a long time."

That Velikovsky proposed both a rain of hydrocarbons and a later precipitation of carbohydrates has been taken by many – notably Carl Sagan – to imply that Velikovsky didn't know the difference between the two (hydrocarbons are typified by petroleum, carbohydrates by starch); and throughout Velikovsky's *Worlds in Collision* (1950) there is a sneaking suspicion that such critics may be right. However, in this specific instance the case must be considered unproven: the text is just not clear. However, the *assumed* confusion has caused later supporters of Velikovsky to try to devise a mechanism whereby hydrocarbons could transmute into carbohydrates: "The hydrocarbons in the comet's tail that had saturated the earth with burning naphtha and drenched it in petroleum were now being slowly changed within the earth's atmosphere, possibly by bacterial action, possibly by incessant electrical discharges, into an edible substance . . ." writes Fred Warshovsky in *Doomsday: The Science of Catastrophe* (1977).

The comet was to return later, narrowly missing the Earth and temporarily halting the Earth's rotation for the benefit of Joshua, before passing onwards to almost collide with Mars. This threw Mars out of its orbit, so that it in turn twice narrowly avoided collision with the Earth. Eventually Mars and the comet (Venus) adopted their present orbits and the Earth was left in peace – and, oddly, with orbit, rotation and orientation almost identical to those it had had before.

Velikovsky was not the first to suggest close collisions with comets had played a part in Earth's – and, more specifically, human – history. Similar theories had been produced by William Whiston (1667–1752), writing in 1696, and by Ignatius Donnelly (1831–1901), of Atlantis renown (see page 105), in his book *Ragnarok* (1882).

After the publication of *Worlds in Collision* (1950; earlier he had written *Ages in Chaos*, unpublished until 1952), Velikovsky was compared favourably in the press with Newton and Einstein. The reason is obvious. The book is full of detailed citations and apparently rigorous correlations between diverse ancient sources. This mixture of ancient legend and celestial dynamics was at the time rare. Moreover, astronomers liked the archaeology while archaeologists liked the astronomy. Unfortunately, the corollary was true, too.

The scientific community reacted with fury to the book – even before it was published. Soon after publication, while the book was in the bestseller lists, publication rights in the US were transferred from Macmillan to Doubleday, because scientists threatened to boycott the former's academic textbooks. To the public, this storm of hysterical protest appeared much like the (assumed) reaction of the Vatican to Galileo's Copernicanism. Looking back, it seems scientists were simply unable to formulate an appropriate response to the 20th century's first major work of pseudoscience.

The theory has imperfections. There is no relevant reference in the oriental astronomical records, which go back for the necessary millennia; one would expect at least a passing mention of such a dramatic event. Moreover, there is on record the Babylonian conceptual breakthrough in realizing that Venus is not two objects (morning and evening stars) but one: this was around 1600BC, while Moses was supposed to be dodging the comet much later than that, some time around 1200BC. It would seem almost boorish to point out that Venus's mass is vastly greater than that of any comet. And one significant problem with the notion that Venus, whatever its mass, swooped close by the Earth – as near as 1000km if the Velikovskian evidence is to be internally consistent – is that the Moon is still in its orbit. If Venus had indeed passed between Earth and Moon, the almost inevitable consequence (aside from all the other consequences) is that the Moon would have been hurled off into space; it is almost inconceivable that there could have been any other result as two large bodies and one small one interacted gravitationally. *Almost* inconceivable: there's the remotest possibility that the Moon could have been put into an extremely elliptical orbit, and that this orbit would have retained something like an orthodox plane. However, the Moon's orbit is now only very slightly elliptical. Well, the highly elliptical orbit could have settled down over time. Alas for this line of reasoning, the Hebrew calendar, drawn up nearly 2500 years ago, shows that the month was almost exactly the same length then as it is now (the month is *very* slowly lengthening as the Moon *very* slowly recedes from us). According to Velikovsky's timetable, this calendar was drawn up less than 1000 years after Venus's close encounter; while the Moon might have eventually recovered a stable, not-highly-elliptical orbit, this would have taken a matter of many, many thousands of years.

And a final point: had there been such a close flyby of Venus, the earthquakes, volcanism and tides would have together conspired to extinguish all life on Earth down to at least the unicellular level.

For a full account of the hypotheses and the problems – such as the bizarre dynamics required for Venus to behave in the manner specified by Velikovsky, the difficulties inherent in any theory which seeks to trace the origins of comets to volcanoes on the major planets, and so forth – *Scientists Confront Velikovsky* (1977), edited by Donald Goldsmith, is strongly recommended. Unfortunately, Velikovsky withheld from that book the paper which he himself presented to the conference.

THE LUMINIFEROUS AETHER

The tale of the luminiferous aether, that mysterious undetectable substance that permeated all space and served as the medium through which light and heat travelled, goes back as far as the 17th century and the debate between the wave and corpuscular theories of light. Christiaan Huygens (1629–1693), for example, thought of light waves as being compressional, like those of sound: as a guitar string vibrates, it alternately compresses and decompresses the air around it in all directions; these regions of compression and decompression are transmitted through the air to vibrate our eardrums, and we hear the pattern of compressions as a sound. But, if light waves were of this same compressional type, then clearly there had to be some medium for them to compress. It couldn't be air, because of course we see the Sun even though it is separated from us by the gulf of space. Thus the medium involved must be some universal, all-pervading and undetectable "stuff".

Isaac Newton's dogmatism about his corpuscular theory of light – that it comprised a stream of particles, not waves – kept the aether out of physical discussion all through the 18th century. But in 1801 the UK physicist Thomas Young (1773–1829) performed experiments which reinstated the wave theory. By the latter part of the 19th century there was, too, a prevailing feeling that *all* physical phenomena could be explained in mechanical terms. The time was right for the reappearance of the aether – which could "carry" not only light waves but also items such as the gravitational and magnetic attractions. Two great supporters of the hypothesis were possibly the two most important of 19th-century physicists: Baron Kelvin (1824–1907) and James Clerk Maxwell (1831–1879).

Light, of course, is not made up of compressional waves –

as was by now fully appreciated; but nevertheless a light wave travelling through a medium such as the aether should cause the build-up of an accompanying compressional wave. As there was no sign of this ancillary wave, the conclusion had to be that the aether was not compressible – i.e., it was *solid*! And this was only one of its odd properties. The high velocity of light showed the aether was exceptionally elastic and yet of minimal density. It was very strong, too: the gravitational force holding the Earth to the Sun was, according to one source, equivalent to one million million solid-steel pillars each 10 metres in diameter; allowing for the spaces between the pillars, this array would need as its base a square approaching 10,000km to a side. It must also be frictionless, since fast-moving objects like planets showed no signs of being slowed down by it.

But surely, surely the aether must be detectable! The German–US physicist Albert Michelson (1852–1931) contrived a clever experiment. As the Earth rushed through the aether, a ray of light sent out at right angles to the Earth's direction, then back to the starting point, would always arrive home earlier than one sent in the direction of the Earth's motion and reflected back from "upstream". Michelson tried several times to detect this travel-time difference, his most sophisticated experiment being that done in conjunction with US physicist Edward Morley (1838–1923), in 1887. No difference in travel times could be detected.

This was a shock. There seemed to be only two possible explanations:

(1) The Earth was at rest in the aether. Somehow, though, this smacked of geocentric cosmology – why should the Earth, of all the objects in the Universe, be the only one at "absolute rest"?

(2) The suggestion put forward by Irish physicist George FitzGerald (1851–1901) and by Dutch physicist Hendrik Lorentz (1853–1928) was valid, however extraordinary: that moving objects become slightly shorter in the direction of their motion (thus the "measuring rods" changed their length in such a way as to conceal any discrepancy in the rate of travel).

Possibility (3), that the aether did not exist, seems to have occurred to few apart from Ernst Mach (1838–1916) until Albert Einstein (1879–1955) put forward his Special Theory of Relativity in 1905. Relativity scrapped such ideas as "absolute rest" and "absolute motion", and showed that such phenomena as the Lorentz–FitzGerald contraction took place without the necessity of invoking such a dubious quantity as the aether.

THE COSMOLOGICAL CONSTANT

When Albert Einstein was working on his General Theory of Relativity in 1917 he spotted an odd thing: his equations seemed to show that the Universe was expanding. But all the cosmologists maintained that the infinite Universe was, overall, static. Anxious to avoid conflict with "fact", Einstein sneaked into his equations a fudge factor, the Cosmological Constant, which allowed them to deal with a static Universe. Later in life he told George Gamow (1904–1968) that he regarded this as the biggest blunder he had ever made.

Not long before Einstein published the General Theory, the US astronomer Vesto Melvin Slipher (1875–1969) had observed the redshifts of the spiral nebulae; at the time, however, no one knew that the spiral nebulae were in fact distant galaxies – this was a discovery that would have to wait until 1923, when Edwin Hubble (1889–1953) spotted Cepheid variables within the nebulae, and thus realized the nebulae must be very large, very distant assemblages of stars, not relatively nearby gas clouds – and so Slipher's observation did not seem to affect the consensus view of the Universe as static: there were other plausible explanations for the redshifts than that they were due to the Doppler Effect (i.e., were receding). So Einstein's assumption of a static Universe was a reasonable one for his time.

Curiously, later in the same year that Einstein published the General Theory, the Dutch astronomer Willem de Sitter (1872–1934) published an alternative solution to the field equations of the theory. The fact that de Sitter's solution seemed to predict a Universe that was *not* static was regarded at the time as being a major weakness in the solution. Even after Hubble's 1923 discovery of the true nature of the spiral nebulae, it was not immediately evident that the Universe was expanding. Although during the 1920s Arthur Eddington (1882–1944), Alexander Friedmann (1888–1925), Georges Lemaître (1894–1966) and a few others proposed this indeed to be the case, they were generally ignored. It was only in 1929, when Hubble and Milton Humason (1891–1972) showed that the redshifts of the galaxies increased with their distance from us, that the expansion of the Universe became fully accepted. Einstein's original equations had been perfectly correct, and the Cosmological Constant shuffled off the stage – apparently to Einstein's great relief, because he almost obsessionally disliked extra complications cluttering up the elegance of mathematical solutions.

Oddly enough, while Einstein's "blunder" in introducing the Cosmological Constant is perfectly forgivable given that everyone thought the Universe was static – a theory can only be as good as the observations upon which it is based – he *did* make a blunder in connection with it. He failed to realize that, even with the Cosmological Constant, his equations could allow the Universe to be static only under very specific circumstances, circumstances that could not last long: any slight variation in them and the Universe would inexorably start to expand. Steven Weinberg (b1933) has suggested that possibly Einstein's rue over his "greatest blunder" was because he had been just a hairsbreadth away from correctly predicting the expansion of the Universe.

In more recent years, with the discovery that the expansion of the Universe is not the simple affair it was once thought to be, there have been claims that perhaps Einstein was right after all to have introduced the Cosmological Constant. The Constant was a repulsive force that, unlike the attractive force of gravity, increases with distance. This would make it seem roughly equivalent to the modern cosmological concept of dark energy. But to equate Einstein's introduction of the

Cosmological Constant with a prediction by him of dark energy is somewhat hagiographical: the resemblance between the two concepts is more seeming than real.

CONTINUOUS CREATION

As noted, the redshift in the light we see from the distant galaxies is a result of the fact that they are receding from us. Moreover, the further away they are, the greater the redshift and hence the greater the velocity with which they are receding. This does not imply that we are at the centre of the Universe. Imagine a balloon on whose surface you have painted pictures of the galaxies; now imagine that you blow up the balloon. As you do so, from the point of view of any one of your painted galaxies *all* the other galaxies are receding – and the furthest ones are receding most rapidly.

Because of this, it has for some decades been assumed that, in the distant past, all the matter/energy of the Universe was concentrated in one spot: the "cosmic egg" exploded perhaps 15 billion years ago, and since this Big Bang the various bits of matter – galaxies – in the Universe have all been receding from each other.

But suppose things were not that simple. Suppose the Universe never had a moment of creation, that it is eternal. Then suppose that, at any point during the infinite history of the Universe, wherever and whenever you were, you would still see the other galaxies receding from you. Such a Universe, infinite in terms of both space and time, was described in the Steady State hypothesis proposed by Thomas Gold (1920–2004), Hermann Bondi (1919–2005) and Fred Hoyle (1915–2001) in the late 1940s. They suggested the observed recession of the galaxies was due to the "continuous creation" of matter from nothing. This idea seems to be in direct contradiction with common sense – but then so, surely, is the idea of the Big Bang.

According to the Steady State hypothesis, as atoms of matter plop into spacetime, each minutely increases the overall "size" of the fabric of spacetime: the net effect is that the galaxies are "pushed" apart – the new matter, of course, being used eventually for the formation of new galaxies. That is, space itself is *stretched* in order to accommodate the freshly arrived matter. Since the infinite Universe is of infinite mass, it is hardly meaningful to talk of the overall mass of the Universe increasing as a result of this process. In addition, the rate of appearance of new atoms does not need to be very high in order to account for the observed recession of the galaxies. In *Frontiers of Astronomy* (1955) Hoyle said: ". . . one hydrogen atom must originate every second in a cube with a 160 kilometre side: or stated somewhat differently . . . about one atom every century in a volume equal to the Empire State Building."

The theory fell. When one looks through a telescope at distant objects one is looking backwards in time. If the Steady State hypothesis were correct, distant parts of the Universe would look much like the nearer parts; and unfortunately this is not the case. And so we can forget the notion of matter popping into existence out of nowhere.

Or can we? It is theoretically possible that *pairs* of fundamental particles may indeed appear from nowhere: each pair is composed of one particle of matter and another of antimatter. In the normal course of events, the two collide and annihilate each other almost immediately, so that the pair-creation makes no overall difference to the Universe – which is exactly why the process can occur. But imagine that the particular pair-creation occurred near to a strong source of gravitation – e.g., a black hole. Then, possibly, while one particle plunged to oblivion in the black hole, the other might shoot off to join the rest of the Universe. In other words, matter may indeed be regularly appearing "out of nowhere", despite the strictures of common sense.

There's a further – and more dramatic – context in which continuous creation is feasible. Recent advances in the mathematics of string theory (or M-theory, as it is now more generally known) have suggested that in a curious way Hoyle and the others may have been right after all – but on a scale far grander than any they conceived. It has long been a mystery as to why the Universe seems to have been so finely tailored to favour the emergence of carbon-based life: had just one of its fundamental properties been only slightly different, the Universe would necessarily be sterile – or, at least, devoid of life as we know it. The obvious response is that, had the Universe been lifeless, we wouldn't be here to wonder why it was so; but this, known as the "weak anthropic principle", is not much liked by scientists, on philosophical grounds. M-theory suggests that the possible number of stable sets of physical laws the Universe could have had is enormously larger than we thought, making the "coincidence" that the Universe should be suitable for us to live in a near-infinitely longer shot. Turning this around, however, reveals the possibility that the Big Bang which brought our own Universe into existence is only one in a hugely large – perhaps infinitely large – ongoing sequence of Big Bangs occurring in different, mutually inaccessible regions of some broader version of spacetime. In such a scenario – a multiverse – the emergence of one universe, or several universes, capable of fostering life does not seem like such a coincidence after all; indeed, it's an inevitability. But a constant sequence of Big Bangs is, in fact, continuous creation. The difference between this concept and that of Hoyle, Bondi and Gold is in essence merely one of scale: Hoyle talked of the need for the creation of just one atom each century in each volume of space the size of the Empire State Building. The truth might be that each century is seeing the birth of, not just widely scattered atoms, but a multiplicity of universes.

STRANGE COSMOLOGIES

Giordano Bruno (1548–1600), burnt at the stake, included a version of Copernicanism as really just an incidental to his own mystical, theistic cosmology; he seems to have despised Copernicus as a mere mathematician, and to have accepted the planets' revolution about the Sun for reasons more associated with magic than with science. Bruno's cosmology is hard for the modern mind to understand, but appears to have had strong connections to animism. The Universe was of infinite extent, and contained an infinite number of inhabited worlds. There was no deity who could be regarded as an individual; instead, the magic of Nature was the deity, present in all things. This deity was reflected in human beings in the form of the creative imagination.

However crazy these ideas might seem if proposed today, Bruno's system would look like almost a paragon of rational scholarship if put alongside some of the more recent pseudoscientific cosmologies.

In the field of unorthodox cosmology there are two main schools. One aims to blast your brains out, whatever the facts; the other seriously asks serious questions, and deserves serious attention: even if the ideas are wrong, one learns something, and the ideas could just be right. It was frequently said of Fred Hoyle, with only minor hyperbole, that he was the single most productive cosmologist in human history because, every time he came out with a new hypothesis, all the other cosmologists immediately stirred themselves into a new frenzy of productivity as they sought to refute him. For some decades, the story of cosmology was thus largely one of "proving Fred wrong".

The Nazis had a perfectly logical reason for believing that Relativity was bunk: *all* "Jew science" was bunk. Hence the Nazis espoused such nonsense as Hanns Hörbiger's World Ice Theory (see page 97). The same excuse cannot be made for many of Einstein's other critics.

A cornerstone of Special Relativity is that nothing can be accelerated up to and beyond the velocity of light. It is this statement which many unorthodox cosmologists have queried.

Charles Hoy Fort posed the most fundamental question: has it yet been proved that light actually has a velocity at all?

A more serious criticism of Relativity was put forward by Herbert Dingle (1890–1978) and is summarized in his *Science at the Crossroads*, (1972). Dingle asked a single question of the "relativistic physicists" and received no satisfactory reply. The question was, in brief:

> If A is moving at great speed relative to B, B will notice that A's clock is running slow (according to Einstein). But, *so far as A is concerned*, it is B who is moving at great speed, and so A will notice that B's clock is running slow. Can two clocks *really* each be running slower than the other?

Muons are extremely short-lived particles (their half-life is about 1.5 millionths of a second). At high altitudes in the atmosphere, muons from cosmic rays are plentiful, generally travelling vertically downwards at a velocity approaching that of light. By taking muon counts at high and at low altitudes one can, therefore, find out if time-dilation is operating for these fast-moving particles (i.e., find out if more of them are reaching the ground than "they have time to"). Such experiments have been carried out, and they yield the expected relativistic results: the muons are surviving far longer than they could if their "clocks" weren't running slow. Put another way, this means that the "average" muon is surviving for a time y rather than a time x, where y is greater than x.

Dingle was saying that it is all very well for us to say the particle is surviving longer than it "ought to"; but consider the matter from the viewpoint of the particle. It sees the Earth rushing towards it at high velocity, and notices that, while so far as it is concerned only a time x passes, clocks on Earth record a time y. That is, the particle thinks Earth's clocks are running *fast*. Help! Relativity tells us quite clearly that the muon should think Earth's clocks are running *slow*.

The answer to Dingle's point is that, because of the velocity with which the Earth is moving towards the muon, the Earth and its frame of reference are shrunk in the direction of motion (from the particle's point of view); in other words, the column of atmosphere through which the particle passes is effectively shorter than it would be if the Earth were moving at a more everyday velocity. From the Earth's point of view, it is of course the muon and *its* frame of reference which are shrunk – with the same result.

To an outside observer, then, either the Earth is getting to the particle more swiftly than it "ought to", or the particle is reaching the Earth similarly too rapidly. Thus time *y* doesn't enter into the matter – which is a good thing, because it was a notional timespan based upon "faulty computation" of the distance involved. So far as our outside observer is concerned, *both particle and planet* have got their clocks "wrong".

This explanation is grossly oversimplified, and should not be taken too literally, but it does indicate the subtle fallacy in Dingle's reasoning: the lesson of Relativity is that distance and time are inseparably intertwined in the form of velocity (even the units of velocity – e.g., kilometres per hour – emphasize this), so that taking one in isolation from the other can lead to confusion.

Dingle was shabbily treated by the scientific community, which seemed to feel that if it ignored him for long enough he would simply go away. His letters on the subject were refused publication in *Nature*, and there seems to have been a degree of petty academic skullduggery at work which reflects little credit on our physicists. This was probably because most scientists did not – and still do not – actually *understand* Relativity, but were content to accept that it "worked": it is a brave man who confesses that he does not understand the cornerstone of his own specialist field.

Though Dingle was a serious questioner of accepted wisdom, it's hard to say the same of most of the other anti-Relativists. Here is a sample from *The Case Against Einstein* (1932) by Arthur Lynch (1861–1934): "The Relativists offer three great verifications of their theory, and they claim that on this basis the whole system is justified. These verifications are, as I shall show, nonexistent." Chris Morgan and David

Langford quote this passage in their book *Facts and Fallacies* (1981), and continue: "In summary [of Lynch's arguments], the shifts of spectral lines produced by the Sun's gravity mean nothing because the Sun is a long way away. The bending of light as it passes the Sun means nothing because not all measurements of the amount of bending agreed exactly. The shifting of the orbit of Mercury means nothing because Einstein fiddled his original equations so as to account for it."

The range of unorthodox cosmologies is vast, running from the simplistic to the incomprehensible; some are actually more difficult to understand than the "complicated" orthodox models they seek to replace. At the simplistic end of the scale, there is great charm in the anonymously published suggestion, dating from 1856, that comets are nothing other than space-travelling volcanoes. And then there is the hypothesis of Osborne Reynolds (1842–1912), from the early 1900s, which sets everything back to front. Pieces of matter are actually bubbles of nothingness, moving in a "space" which is in fact filled with tiny, closely packed invisible spheres. Gravity is not a pull but a push – obviously, the bigger the bubble of nothingness, the more it distorts the matrix of spheres, and consequently the greater its gravitational influence.

That gravity might be not a pull but a push was a notion widely promoted in the 1930s and 1940s by Thomas Graydon: the Moon is trying to fall to Earth but the gravitational field of the Earth is pushing it away. Of course, if you jump off the Empire State Building you notice that there is, apparently, a strong downward pull. Not so. It is simply that, so close to its surface, the Earth cannot push you strongly enough to keep you in orbit.

The cosmology of Alfred Lawson (1869–1954), founder of Lawsonomy, is . . . intriguing. The concepts are rather opaque, so the following must be regarded as something of a guess. Lawson, like Goethe, believed the Earth to be a living organism. More complex are his ideas of the Universe as a whole. There is no such thing as emptiness: what we would consider to be empty space is in fact filled with aethers of various densities. There are no such things as forces per *se*, only the movements of denser substances towards less dense ones, governed by the three concepts suction, pressure and penetrability. Thus there is no force of gravity holding you to the Earth: it is simply the Earth's suction. Your eyes draw light into themselves by this same suction.

Lawsonomy is not just a cosmology but a complete knowledge system. To give some idea of the scope of this metatheory, here is a quote from one of Alfred Lawson's many publications: "If it isn't real; if it isn't truth; if it isn't intelligence; then it isn't Lawsonomy." Everything else is. Because of its extraordinary range, it is almost impossible to actually understand Lawsonomy, let alone summarize it. Matters are not helped by Lawson's astonishing (or possibly self-parodying?) egotism; his comments about himself and his system include "The birth of Lawson was the most momentous occurrence since the birth of mankind" and "In comparison to Lawson's Law of Penetrability and Zig-Zag-and-Swirl movement, Newton's law of gravitation is but a primer lesson, and the lessons of Copernicus and Galileo are but infinitesimal grains of knowledge." To show that he was not a vain man, when writing about himself he usually employed the pseudonym "Cy Q. Faunce".

One charming notion of Lawsonomy is that inside your brain there are microscopic creatures who, essentially, do your thinking for you. There are two types: the organizing menorgs and the disorganizing disorgs. Menorgs seem to act like armies, with generals instructing and coordinating the efforts of large numbers of private soldiers who operate all your brain processes. Against this virtuous rule, the disorgs wage a sort of guerrilla war. (Thomas Alva Edison had a similar theory.)

But Lawson's most important concept was the law of Zig-Zag-and-Swirl. So far as I can understand it, this seems to state in a pompous fashion that nothing moves in a simple path. If I walk across my study to look at a book, I may think I have moved a few metres in a straight line; but the Earth is rotating and travelling around the Sun, which is itself moving in space, etc., and so my *absolute* path has been a very complex one – it has zigzagged and swirled. There is nothing *wrong* with this "law", of course, just that it's a statement of the obvious.

Charles Fort made a staggering number of cosmological suggestions, most of them outrageous and at odds with observed reality, and some of them inconsistent with each other (a mutual inconsistency of which he was aware, and perfectly cheerful about). To list all of them would be impossible: here are just a few. The sky is gelatinous – if not entirely, then at least in parts. Stars are holes in the shell of jelly; they twinkle because the jelly trembles. Floating somewhere above the Earth is an island called Genesistrine. It is invisible, and suffers the nuisance of its contents frequently falling over its edges to rain down on us; this is where all those showers of hazelnuts, artefacts, frogs, cats and dogs come from. The Earth does not rotate. Actually, Fort was prepared to compromise about this: it *might* rotate but, if so, only about once a year. "As I have admitted before I'm intelligent," confessed Fort.

Wilbur Glenn Voliva likewise knew that the Earth – in his case flat – did not rotate: otherwise we'd be able to jump up in the air and land hundreds of kilometres away. His other cosmological ideas have a passing interest, too. The Sun is only about 5000km away, and only about 51km in diameter. This is because the only possible purpose of the Sun must be to illuminate humanity, and there would thus be little point in putting it any further away; if you want to read in bed, you don't erect a searchlight in the next city. The stars, too, are small and quite nearby, while the Moon shines with its own light. Actually, it's difficult to fit the Moon into his cosmology, the Sun being so near.

Towards the end of *Worlds in Collision* (1950) Immanuel Velikovsky hints at his own unorthodox cosmology. He picks up the idea that the atom might be rather like the Solar System, but in microcosm, with the nucleus as analogue of the Sun and electrons as analogues of the planets. He notices that electrons frequently change their energy levels, moving from an inner orbit to an outer one or *vice versa*, and suggests that this is exactly what happened when, according to Velikovskianism,

Venus and Mars danced around the heavens. Earth and Moon, too, have undergone such shifts – odd, since the Moon is not now an independent planet.

One flaw might seem to be that electrons do not orbit the nucleus in the same way that planets orbit a star. Each planet in the Solar System has its own orbit (although Earth and Pluto may be regarded as binary planets), whereas, in the simplest atomic model, each orbit around the nucleus is characteristically occupied by a plurality of electrons. Moreover, all electrons seem to have the same charge (i.e., electromagnetic influence) whereas the planets, being of all different sizes, exercise widely different gravitational influences. Besides, electrons aren't really material in any everyday sense of the word: they're certainly not solid bodies like planets.

In *The Intelligent Universe* (1975) David Foster pointed up the similarities between the Universe and a vast electronic computer, a "giant brain". These are, essentially, that the computer stores data in the form of bits, which can be considered analogous to subatomic particles; that it processes these bits just as subatomic particles are "processed" in chemical transformations; that a computer is programmed, much as an acorn is "programmed" to produce an oak; that the computer combines bits to produce "words", just as atoms combine to form chemical compounds; and that data transmission in both the computer and the Universe is effected electromagnetically.

Not all unorthodox cosmologies are absurd. Take the notion of the Selfish Biocosm. Many thinkers, mystical/religious and otherwise, have speculated about the fact that the Universe and all its various laws and constituents seem so very well tailored to the emergence and development of life, and have used this as an evidence of a purposeful creation. Most scientists point out that this is rather to turn the reality on its head, as if fishes were to imagine the oceans had been tailored to suit their needs rather than the other way around: that evolution has tailored them to fit the marine environment. Most scientists, but not all. James N. Gardner has at least some scientific qualifications, his primary field being complexity theory, yet in his *Biocosm: The New Scientific Theory of Evolution: Intelligent Life is the Architect of the Universe* (2003) he turns the tables with a vengeance on the conventional view.

Gardner's hypothesis is that hardwired into the Universe at the most fundamental level are the conditions that make the emergence and evolution of carbon-based life – "life as we know it" – inevitable. Further, not only does the nature of the Universe itself demand that life spring up prolifically all over the cosmos, it has as an inevitable consequence the development within the myriad lifeforms, and even more so after they come together and interact, of a sum intelligence beyond anything we can so far dream of: the pooled intelligence of all the lifeforms will become, in effect, the brain of the Universe, enabling it to reproduce itself. Individuals and even individual high-technological civilizations can thus be seen as analogues of the genes that both make us and enable us to reproduce and evolve; Gardner's name for his hypothesis is a deliberate reference to the Selfish Gene idea of Richard Dawkins (b1941).

In many ways the Selfish Biocosm, which was in a sense presaged by David Foster's "intelligent Universe" hypothesis, is a startlingly attractive idea. Darwinian evolution on Earth can be seen as a mere component of a colossally larger system. Looked at another way, while the hypothesis would seem to outlaw any possibilities of the Creator beloved of the religious, it also implies that humankind, together with all of the Universe's other intelligent lifeforms, is evolving towards becoming something indistinguishable from God. This does allow a theistic loophole of sorts. If universes spawn "baby" universes, then our Universe is presumably itself an offspring, born in consequence of a "parent" universe having developed within itself the kind of godlike supermind Gardner envisages. That could be close enough to "God created the Universe" for some.

To call the Selfish Biocosm concept a hypothesis is probably to undervalue it: whether or not it proves to have any validity, it is a full-blown theory, seeking to explain a very wide diversity of phenomena on the grandest possible scale – the multiuniversal scale, in fact. Gardner is well aware of its potential as the basis for that Philosophers' Stone of physics, the Grand Universal Theory – the final theory that pulls everything together in one coherent form.

As one might expect, the concept's reception in the scientific community has been mixed. One difficulty is that it is infernally hard to test, certainly within the lifetime of a single human being – or perhaps even within the lifetime of a technological civilization. We may be – in fact, are most likely to be – just fishes marveling at the way the ocean was designed for us. Yet who could deny the appeal of the concept's Stapledonian grandeur?

THE WORLD ICE THEORY

Possibly the best known of all unorthodox cosmologies, the World (or Cosmic) Ice Theory was the brainchild of an Austrian mining engineer Hanns Hörbiger (1860–1931). In 1913 he published a vast tome called *Glazial-Kosmogonie*, which sets out the theory in some detail. The theory is so complex and so radically at odds with common sense that I can do little more than hazard a summary of it here.

The stars are mere chunks of ice. The only body in the Universe that is actually a star (in our sense of the word) is the Sun. Everything in the Universe orbits the Sun. However, since all space is filled with hydrogen, albeit rarefied, friction ("air resistance") causes all orbits to decay: the various celestial objects spiral down towards the Sun, and finally fall in. Whenever this happens, a sunspot is seen. If the body was a star, its ice is vaporized and part of it blasted back into space; by the time this ice reaches the Earth it has refrozen to form tiny crystals: that's where high-altitude clouds come from.

In fact, the Earth and Sun are odd-men-out in the Universe, for all the other celestial objects are, if not entirely of ice, then at least entirely ice-covered. For example, Hörbiger claimed that Mars is covered in an ocean of ice or water some

400km deep, and that Mars would one day become a moon of the Earth (see below) – unless, that is, it "missed" and instead plunged straight into the Sun. In fact, the Earth and all the other planets are destined to meet their dooms by dropping into the Sun, but they will be replaced. Hörbiger seems to have thought there is an infinite number of planets beyond Pluto, and that planets approach the Sun rather as if they were riding on some cosmic conveyor-belt.

Reassuringly, it will be quite some time before our world plummets to its fiery death, but there is a shorter-term problem to worry about: the Moon is spiralling in towards the Earth, and will collide with it. This is not the Earth's first moon: it has had several predecessors, the most recent of which crashed down in comparatively recent prehistoric times. This catastrophe neatly explains various legends – most notably that of the Flood. Similarly, the capture of our present Moon caused various difficulties, such as a nasty dose of Pole Shift and the sinking of Atlantis. Denis Saurat, in *Atlantis and the Giants* (1957), used the theory to explain the reference in *Genesis:* "There were giants in the Earth in those days." As a past moon spiralled in towards the Earth, its gravitational pull counteracted that of the Earth – and hence people could grow very tall.

The World Ice Theory might seem just plain cranky – and no more dangerous than any of the other unorthodox hypotheses described in these pages. But it became inextricably involved with the rising tide of German Nazism – for two main reasons. First, the antisemitic nationalists had a strong urge to reject all physics and cosmology based upon the ideas of a Jew (Einstein): they wanted a new, Aryan, cosmology – and the more different from the "Jewish" one the better. Logic, rationality and all the other intellectual tools were flimsy dams against this flood of hatred. That this new cosmology came from an Austrian amateur scientist was an added advantage: after all, was not Hitler himself an Austrian amateur politician? Second, some theorists suggested that the reason for the

The Solar System with a veil of ice, from the Hörbiger-Archiv, Vienna.

Aryans' natural superiority to all the rest of mankind was a result of the "toughening" of their ancestors in the chilly North during the last glacial age. Here was ice again! It couldn't be coincidence. Obviously, then, to reject the World Ice Theory was to reject Aryan superiority.

A comment from a pamphlet produced in 1953 by the Hörbiger Institute sums it up: "The final proof of the whole cosmic ice theory will be obtained when the first landing on the ice-coated surface of the Moon takes place."

Carnac depicted in 1846 in an engraving by Louis Haghe.

CHAPTER 2

LOST WORLDS, LOST PEOPLE, LOST CREATURES

THE PAST IS A FOREIGN COUNTRY. They do things differently there. Or so proposed L.P. Hartley (1895–1972) in *The Go-Between* (1954). When we are talking of the time before writing was invented we can have little or no conception of quite *how* differently people thought then. The only clues we have are enigmatic artefacts, on the one hand, and imaginative and often lurid mythology, on the other. Interpretations of these generally tell us more about our own thought-processes than about those of our ancestors. Take the case of the Scottish astronomer Charles Piazzi Smyth (1819–1900). When one of the Great Pyramid's casing-stones was dug up from the desert it proved to be just over 63cm across. Smyth decided that 1/25 of this length was a "pyramid inch". The Earth's polar radius is supposed to be exactly ten million pyramid inches – a further "proof" of the validity of his theory. Using the pyramid inch, Smyth made the most amazing deductions about the past, present and future from measurements he took in the Great Pyramid. All the dates tallied, including – or so Smyth assumed – the beginning of the world in 4004BC and its end in either 1882 or 1911. It was a staggering piece of research.

Unfortunately, it never occurred to Smyth that the ancient Egyptians might have been quite happy to make the casing-stones of the Great Pyramid all of different sizes. That was completely contrary to his own Victorian, British sensibilities: in the wake of the Industrial Revolution, everyone expected engineering work to be done to high precision. When further casing-stones were examined, it was discovered that the length of the "pyramid inch" varied according to which casing-stone you chose for your base measurement.

The dimensions and orientation of the Great Pyramid have given rise to a plethora of amateur hypotheses. Most of the theorists say that it is impossible that such a large edifice could have been built to house just a single corpse, and one is almost tempted to believe them: a good case can be made for the Great Pyramid's having been used as an astronomical observatory. Yet, *all over the world*, similar efforts were and often still are made to commemorate the passing of an aristocrat or ruler. Think of the Viking habit of burning a ship as the funeral pyre for a king – at a time when ships represented mighty investments of human energy. In modern days, state funerals are as glorious, and expensive, as ever. Moreover, in Egypt the evolution from the fairly primitive mound-type burial chambers, *mastabas*, to pyramids can clearly be traced.

A book could be filled with unorthodox theories about the Great Pyramid itself; here are a couple of representative ones. Dr George Hunt Williamson (1926–1986) told us that ancient astronauts who came here 18 million years ago built the monument some 24,000 years ago. This should be easy enough to check, since all we have to do is dig up the spaceship which he claimed the aliens buried underneath the Great Pyramid. And there is a persistent rumour that inside the edifice was found a model landscape, through which ran rivers of mercury. When this model was found the rivers were *still running*! It seems a tragedy that this model is, for some strange reason, no longer extant.

The most popular current theory concerns pyramid power, the ability of objects shaped like the Great Pyramid to preserve objects placed within them. Craig and Eric Umland tell us that "a body left for any length of time in a pyramid will *automatically become mummified*" (their italics). But, you ask, why then go to all the bother of extracting brains, individually wrapping soft organs, embedding the body in wax, and so on? "Undoubtedly, the mummies which have been found were late-comers, interlopers." So it worked for the original alien corpses, which have long vanished, but not for humans?

The modern popularity of pyramid power is largely due to Lyall Watson (b1939), who in *Supernature* (1973) told of a Monsieur Bovis, who found that cardboard pyramids were useful for mummifying dead cats, and of Karel Drbal, a Czech radio engineer who in 1959 patented the model pyramid as a device for keeping razor blades sharp. Watson himself claimed to have used a pyramid to keep his own blades sharp for up to four months. Watson did not cite his source for his tales of Bovis and Drbal, but it seems to have been *Pyramid Power* (1973), by G. Patrick Flanagan, a Californian whose thriving business sells . . . model pyramids.

Watson, in the matter of razor-blade preservation, challenged: "Try it yourself." Unfortunately, this researcher does not shave, and so resorted to putting some chewing-gum, rather than a razor blade, in a correctly oriented model pyramid to see if it lost its flavour. It did.

ATLANTIS

The Land of the Amazons, Arcadia, Atland, Atlantis, Avalon, the Isles of the Blest, St Brendan's Isle, Brittia, the Cassiterides, the Land of El Dorado, Eden, Faeryland, the Fortunate Islands, the Gardens of the Hesperides, Hyperborea, Ierne, Lemuria, Lyonesse, Mayda, Mu, the Land of Prester John, Shambhala, Shangri-La, the Land of the Queen of Sheba, (Ultima) Thule, Troy, Uranus (not the planet), Ys . . . a listing of lost lands and continents could extend forever. In addition, we should include also such places as Mars, the Fifth Planet and Venus, all of which fall into the same category.

Is there any truth at all in such tales, or are they total fantasies? Atlantis could have been Crete, Thule could have been the Orkney Islands, or Iceland, or . . . Yes, just possibly, some of the lost lands could really have existed. But somewhere

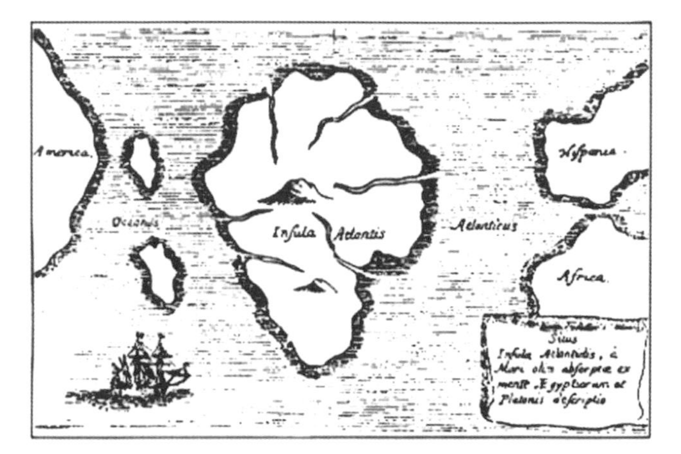

Atlantis mapped in the 17th century by Athanasius Kircher, better known for his work in microscopy (see page 279).

amongst all the legends there ought to be just *one* about a land which was not heavenly, and whose inhabitants were not in some way superhuman. Where, in short, are the sewers of Atlantis?

Not all of the lost lands were lost in ancient times; some were last seen relatively recently. Take the case of the Aurora Islands, an archipelago in the South Atlantic, to the southeast of the Falklands. These islands were first spotted in 1762 by the crew of a ship called the *Aurora* – hence their name. In the 1790s they were seen by two Spanish ships, who this time measured the archipelago's latitude and longitude. Despite never being seen again, the Aurora Islands appeared on maps for a further century or so. It's feasible what the Spaniards saw were icebergs – feasible but unlikely, because the seamen would surely soon have realized their mistake. If we discount hoax or gross navigational error, the matter remains a mystery.

Best known of all the lost lands is, of course, Atlantis. The main Atlantis legend is well known. It seems to have started its life in the 4th century BC, when Plato wrote in the *Critias* and *Timaeus* that Solon (*c*638–558BC) had been told by an Egyptian priest of documentary accounts of a now-vanished nation beyond the Pillars of Hercules. Atlantean scholars have made much of the fact that Plato, in *Timaeus*, seems to make a point of declaring (through Critias) that the story of Atlantis is a true one. The difficulty with this argument is that in almost *all* his dialogues, no matter how obviously fanciful, Plato stated that the stories were true: he was relying on his educated audience to realize that of course they weren't – they were just parables introduced for the purpose of conveying a moral or encouraging philosophical thought. In short, Plato's statement of truth, delivered through the dialogue of the fictional Critias, is merely a literary device.

Even so, Atlantis has been popular almost ever since, most especially after Ignatius Donnelly (1831–1901) published his *Atlantis, The Antediluvian World* (1882). The Theosophists leapt aboard this bandwagon with their customary alacrity. It is now generally thought that the Atlantis legend, if it has any factual basis at all, refers to the destruction in about 1400BC of the Minoan (Cretan) civilization, when the volcanic island of Thera (Santorini) erupted with a violence beside which that of the 1883 Krakatoa eruption pales.

Donnelly described Atlantis as the home of the original Aryans, who, despite having either invoked the wrath of God by loose living or destroyed their homeland through waging war with fearsome weapons, were still much better than the degenerate other races (us). This was picked up in the 1920s by Karl Georg Zschaetzsch (b1870): in Atlantis (Eden) dwelt the Aryan master-race, vegetarians all, until a female non-Aryan (Eve) imported the demon alcohol in the form of cider (apples). Unused to such heady stuff, the Aryans went on a master-binge (i.e., were corrupted); but their bacchanal did not last long. Atlantis was destroyed as the Earth collided with the tail of a comet. There were only three survivors of Atlantis's sinking, in Zschaetzsch's Wagnerian pseudohistory. These were Wotan, his sister (who was pregnant) and his daughter (who was not). Wotan's sister died giving birth to a son, who was fortunately suckled by a passing she-wolf. This noble Aryan family, anxious to avoid incest, began eventually to interbreed with the revolting primitives on the mainland. This is why their descendants indulge in such disgusting practices as drinking alcohol and eating meat. All this stuff about the original Aryan master-race naturally appealed to the Nazis and the Theosophists.

The great boom in studies of Atlantis (or Atlan or Atalantica, as it is sometimes also called) began in the latter quarter of the 20th century and is continuing to this day, although the craze reached a fairly major peak towards the end of the 19th century, after the publication of Donnelly's book. W.E. Gladstone (1809–1898) went so far as to request funds from the British Treasury to mount an expedition in search of the lost continent. Rather more recently Dr Paul Schliemann, supposed grandson of the discoverer of Troy, Heinrich Schliemann (1822–1890), claimed to have found among some of his purported grandfather's papers an account of the discovery in the ruins of Troy of a vase inscribed: "From the King Cronos of Atlantis." Unfortunately this document has never been made public.

Founded in 1957 was the Atlantean Society, and its early meetings in London attracted enormous support. The audiences were treated to utterances from a high priestess of lost Atlantis, Helio-Arcanophus, which utterances arrived via the lips of ex-actress Jacqueline Murray. According to Helio-Arcanophus, Atlantean civilization arose as a result of a benevolent experiment by Venusians in an attempt to spur the evolution of our primitive ancestors. Briefly: Venusian souls were put into caveman bodies, and the highly advanced civilization of Atlantis was the result. The Society's magazine, *The Atlantean*, contained many useful and instructive articles by both humans and nonhumans – and provided answers to such urgent questions as "Will spacemen landing on the Moon go mad?"

Not all Atlantean studies are spiritually or psychically based. In 1968 an aerial survey off the coast of Bimini revealed what looked suspiciously like underwater roads and buildings, and the excitement of Atlantean scholars rose once more. Divers were sent to the area, and their first reports seemed to justify the excitement: there were paved roads and tumbled architectural columns . . . not to mention, if we're to believe many of the secondary accounts, great walls, pyramids, and

Stonehenge-like megaliths. Unfortunately, the cylindrical masonry chunks that were thought to be fallen columns proved to be made of concrete, and concrete of a type manufactured from the 19th century onward; it's speculated that what may have happened is that wooden barrels of the dry concrete were being transported by a ship that ran into trouble and either sank or whose crew threw the heavy barrels overboard. The wood rotted away, but not before the now-wet concrete hardened inside the barrels. Even if this speculation proves fanciful, there's no getting away from the fact that the "columns" are concrete and of 19th-century or later date. The "paved roadways" are naturally occurring limestone formations.

Maxine Asher (b1931), Head of the Ancient Mediterranean Research Association at Popperdine University, California, who led an expedition to try to find the continent in 1973, said of the reasons for Atlantis's sinking:

> Atlantis probably went down originally as a result of seismic upheavals under the ocean, but I do believe the final destruction was probably cosmically ordained. Hypothesizing, of course, the final destruction of Atlantis could have come cosmically, because the people had become so evil and had generated enough negative force that they had disrupted the Cosmos. Divine Retribution could have been the springboard – let me call it that – for the cosmic upheaval that destroyed Atlantis.

She believed also that Atlantis extended from the Canaries to Bimini and from Ireland to Newfoundland; that the Atlanteans had a power source as powerful as nuclear energy but "psychic in nature"; and that not one Noah but twelve survived the inundation, the twelve Noahs corresponding to the Twelve Tribes of Israel.

What are we to make of the translation done in 1864 by the Abbé Charles-Étienne Brasseur de Bourbourg (1814–1874) of the Mayan book called the Troano Codex? In this rendition, we discover more of the "true history" of Atlantis. For his translation, de Bourbourg utilized the researches of Diego de Landa Calderón (1524–1579), Bishop of Yucatán, who demonstrated to his own satisfaction that the hitherto-undeciphered written language of the Maya was alphabetical in nature (i.e., it had letters representing individual sounds, as English does) and, further, ascribed to the Mayan symbols their equivalents in Spanish. Unfortunately, the Mayan language is not alphabetical at all – it's hieroglyphic – so de Landa's exercise was one of pure self-delusion and De Bourbourg's "translation" necessarily a work of fantasy. The same can be said of the "translation" by Augustus Le Plongeon (1826–1908) of supposedly the same work, as *Queen Moo and the Egyptian Sphinx* (1896); in this instance the Maya are of ancient Egyptian origin. (Le Plongeon also introduced the concept of Mu – see page 112.)

In his 1992 book *The Flood from Heaven: Deciphering the Atlantis Legend*, Eberhard Zangger makes a plausible case for the legend having been derived from the Egyptian account of the Trojan War, which naturally enough differed quite considerably from that of Homer – and differed even more so by the time it had passed through several hands from Egyptian hieroglyph to Plato. This would tie in with Plato's statement that the tale was originally told in Saïs – then the capital of Egypt – by a priest there to Plato's ancestor Solon (*c*640–559BC), who was visiting. Not realizing that the conflict involved was the Trojan War, Solon started writing down the tale as it had been related to him, but for some reason never completed the manuscript, which was passed down through the family until it reached Plato. Plato in turn incorporated the tale into the two Dialogues, believing it was indeed about a lost land, but suddenly, midway through *Critias*, recognized what he was writing about, and abandoned the effort. Zangger's interpretation depends upon the assumption – now widely held – that Troy was not so much a city as a nation; his hypothesis, based in archaeology and the antithesis of the more fantasticated "explanations" that have seized the popular imagination, is appealing.

In 1977 Robert J. Scrutton announced that there had coexisted with Atlantis *another* Atlantis, called Atland, which lay between the north of Britain and Greenland; Iceland is presumably a remnant. Atland survived until 2193BC, and was then abruptly destroyed, either when an asteroid hit the Earth or as a result of a close encounter with the comet that, according to Velikovsky (see page 78), would one day become the planet Venus. Scrutton's hypothesis seems to depend upon what is generally regarded as a 19th-century forgery, the "Oera Linda Book".

The Scottish mythologist Lewis Spence (1874–1955) likewise added a further lost Atlantic continent to Atlantis. In *The History of Atlantis* (1926) Spence told us of Antilla, and claimed the cultures of the two continents had suffered several major tectonic upheavals through their long history. The refugees from one of these, 25,000 years ago, arrived in Europe and settled there. We now know them as Cro-Magnon Man.

It wouldn't do to leave Atlantis without mentioning that there are still first-generation survivors among us – well, depending on your definition of the word "survivor". Most famous is the 35,000-year-old Lemurian-born Atlantean warrior Ramtha, who, although currently resident in Heaven, is able to channel himself through a woman called J.Z. Knight (b1946); she first heard from him in 1977 while in her kitchen in Tacoma, Washington State, when she chanced to put a toy pyramid on her head. Apparently Ramtha instructed Knight fairly early on that it was right for him to require a donation from those who sought his wisdom; since Knight is his trustee here on Earth, the accumulated donations of the past 30 years or so have made a significant financial contribution to her spiritual well-being. That wisdom covers the various fields of science Ramtha understands far better than modern scientists do – cosmology, quantum physics, neuroscience and archaeology among them. To be sure, modern scientists have difficulty finding any evidence for the various scientific insights that Ramtha offers, but this is their fault, not his: he promotes a non-evidential form of scientific research that can probably best be summed up as: "If it feels good, believe it." There is no objective reality, only subjective realities, so objectivity – such as orthodox scientific research – is necessarily a meaningless exercise.

Ramtha has become sufficiently wealthy and/or influential that in 2004 he was able to release a feature movie called, variously, *What the #$*! Do We Know?* and *What the (Bleep) Do We Know?* The writers, directors and producers were all members of the Ramtha School of Enlightenment, as were some of the actors. In *What the #$*!*, the story of a young wedding photographer seeking help for her depression from various real-life pseudoscientists (played by themselves) is interwoven with clips of scientists of varying degrees of credibility talking about things like quantum mechanics, which the movie then relates to exotic spirituality. Richard Dawkins (b1941) summed up the movie's theme thus: "Quantum physics is deeply mysterious and incomprehensible. Eastern spirituality is deeply mysterious and incomprehensible. Therefore they must be saying the same thing." The movie's moral appears to be that, since everything is subjective, if you can alter your perception of reality you can, by altering reality itself, cure yourself of whatever it was you were feeling fed up about – because, of course, that source of depression no longer exists in the new reality. Part of the woman's depression cure involves drawing little glitter hearts all over her body to symbolize that she accepts herself.

One of the pseudoscientists who appears in the movie is worthy of separate attention. He is Masaru Emoto (b1943), and his experiments – details of which have, alas, yet to appear in any of the scientific journals, although you can read about them in Emoto's book *Messages from Water* (2 vols, 1999 and 2001) – indicate quite conclusively that the power of human thought is sufficient to alter things in the physical world around us. Furthermore, water is capable of understanding spoken or written words, and of responding to them with messages of great wisdom. Emoto's means of demonstrating the wisdom of water is to tape slips of paper containing various words to jars of water. Left overnight, the jars that bear words relevant to pleasing emotions – "love", "beauty", etc. – are found to have developed beautiful crystals. By observing the crystals, you can tell what the water is thinking.

Returning to Ramtha, he has apparently allowed himself to be copyrighted by Knight. In 1992 she discovered that one Judith Ravell, a German, was claiming likewise to have made contact with the famed Atlantean sage. Knight took her to court, and after a three-year legal fight won the case: Ramtha was exclusively Knight's property. In 1997 the Austrian Supreme Court upheld the lower court's ruling.

LEMURIA AND MU

In 1952 George Adamski (1891–1965) met and spoke with a friendly Venusian in the desert. The alien fortunately left footprints in the sands, which George Hunt Williamson was able to examine. He found that the soles of the visitor's shoes had – in place of the usual corrugations – cryptic messages. Williamson was able to translate these, and found that they told us Lemuria would soon reemerge.

Who "invented" Lemuria? According to one version, it was the German naturalist Ernst Heinrich Haeckel (1834–1919), best known for his theory that ontogeny recapitulates phylogeny (see page 142). He was concerned to account for the modern distribution of lemurs, and introduced the notion that there might have been a landmass in the Indian Ocean over which they could have migrated. Haeckel went on to speculate that the ancient Aryan race must have come originally from Lemuria before settling in Asia; so confident was he of this that in his *The History of Creation* (1879) he published a map of the migratory routes he believed the Lemurians/Aryans had followed. An early staunch supporter of the Lemuria hypothesis was Alfred Russel Wallace (1823–1913), Darwin's colleague in elucidating the theory of evolution by natural selection.

But the UK geologist Philip Lutley Sclater (1829–1913) may have beaten Haeckel to it, proposing the existence of Lemuria in 1855 – again to account for the distribution of lemurs. The assumed behaviour of lemmings had convinced Sclater that such lost lands existed: the lemmings were, of course, trying to migrate to Atlantis!

The Theosophists latched onto the idea of the lost continent. According to Helena Blavatsky (1831–1891), the Lemurians were giant, ape-like hermaphrodites who, as time went on, grew more and more to resemble contemporary human beings. The destruction of Lemuria was somehow linked to its inhabitants' discovery of sex; fortunately a few of them had migrated to Atlantis before the catastrophe occurred. Rudolf Steiner (1861–1925) added the intriguing details that the Lemurians had no powers of reason, but used instinct in its place, and that they possessed telepathic and psychokinetic abilities. According to another hypothesis, Lemuria was originally populated 18 million years ago by Venusians. At first, these beings were aethereal, but as time went on they were corrupted by the ghastly Earth environment and became more material. The final straw came when, as per Blavatsky, they differentiated into male and female and discovered sex: the continent was soon destroyed.

In some versions, surviving Lemurians are said to be living hidden from human society as troglodytes. Certain of the (US) Rosicrucian teachings tell us that these subterranean people dwell around Mount Shasta, California.

Lemuria is often incorrectly identified with Mu. The continent of Mu seems to have been invented by Augustus Le Plongeon for his *Queen Moo and the Egyptian Sphynx* (1896), based on Mayan writings of dubious authenticity; as with so many of these "mysterious old documents", no one else was allowed to see them. Mu lay in the Pacific, unlike Lemuria, which lay in the Indian Ocean.

Although it was Le Plongeon's creation, Mu really "belongs" to Colonel James Churchward (1852–1936). Churchward was fortunate enough to find, in various (unidentified) Indian (or perhaps Tibetan) monasteries, a large number of tablets in an incomprehensible language, which Churchward was able to identify as Naacal, humankind's first language. This tongue was understood only by two (unidentified) Indian (or perhaps Tibetan) mystics, who were good enough to give him a crash course deciphering it. He was then able to reveal that Mu had sunk into the Pacific 12,000 years ago, almost all of its 64 million inhabitants perishing, although a few were able to survive on the Pacific islands; from these degenerates (often insane, often cannibalistic) sprang *Homo sapiens*. Today's Aryans are of course physically and mentally the closest to the superbeings of Mu.

Churchward's books, published long after their writing, are *The Lost Continent of Mu* (1926), *The Children of Mu* (1931), *The Sacred Symbols of Mu* (1933) and *Cosmic Forces of Mu* (1934).

THE AMAZONS

The Greek myth of a warrior race of women may have originated as a "mirror story" – a tale of a far-off society in which everything is topsy-turvy (like Samuel Butler's *Erewhon* [1872]). On the other hand, the myth may possibly have a vague basis in fact, inasfar as it was not unknown for the women of Greece's foes to fight alongside their men, *in extremis*.

As Greek civilization expanded its sphere of influence, so the land of the Amazons became further and further away. At one time, indeed, there were thought to be three distinct races of Amazons in Asia, Africa and the kingdom of the Scyths. Nevertheless, it is on record that the Amazons fought boldly in the Trojan War – on the side of Troy.

Did they cross the Atlantic to do so? The Amazon river is so-called because of a report from Francisco de Orellana (*c*1500–*c*1549) that, just prior to his epoch-making descent of it (1540–41), he was told of a nearby race of warrior women.

A mounted Amazon depicted on a Greek vase dating from the 4th century BC.

Contemporary accounts are maddeningly vague, but the land of the Amazons seems to have been Paraguay, or a part of that country. And there may be some truth in the tale. Here is a snatch from J.M. Cohen's 1968 translation of a 1555 account by Augustin de Zárate:

> And Leuchengorma's subjects [Leuchengorma was a local lord in southern Chile] told the Spaniards that fifty leagues further on there is a great province between two rivers entirely populated by women, who will only allow men to come near them at the times most suitable for conception; and if they bear sons they send them to their fathers, if daughters they bring them up themselves . . . [T]heir queen is called Gaboimilla, which means in their language "golden sky", because great quantities of gold are said to be mined in that land; and they make very fine cloth and pay tribute of all their commodities to Leuchengorma.

This may just have been mythology in the making, as two cultures cross-fertilized for the first time, but it has a smack of authenticity. Paraguay is more like 500 than 50 leagues from southern Chile; but it does lie between two mighty rivers, the Amazon and the Paraná.

But back to the Greeks. Reproduction posed a pretty problem for the Amazons since they killed all their male babies at birth. So the myth was embroidered. Military incursions were led by Theseus, Bellerophon and Hercules (among others) and the inevitable happened: what we might primly describe as a cultural cross-fertilization.

The name "Amazon", "breastless", reflects the warriors' pragmatic habit of burning off their right breast (or left, for lefthanders) in order to avoid painful accidents with bowstrings.

THE MOUNDBUILDERS

A far more recent myth concerns the ancient North American race called the Moundbuilders. This fallacy, although roundly debunked by the end of the 19th century, still had considerable sway through much of the first half of the 20th, and still re-emerges occasionally today.

As the White Man spread across North America he came across many examples of settlements that were far in advance of the gathering of tepees stereotyped in Western movies. Sites such as Cahokia in Illinois quite obviously were or had been major population centres, home to thousands of people. They contained edifices that were architectural near-rivals to the Pyramids of Egypt, although sculpted of earth rather than of stone. So impressed was Thomas Jefferson (1743–1826) by mound architecture that he used it as a template when building his house at Poplar Forest, near Lynchburg, Virginia, in 1806. The cultural artefacts associated with these centres – ornaments, ceramics, etc. – were not those of an unenlightened, disorganized nomadic people. Clearly these were the hallmarks of a significant civilization – or civilizations, in the plural. They most certainly did not fit with the image the White Man was so keen to portray of the Native Americans as ignorant, barbaric savages: the Europeans were supposedly bringing civilization to the Americas, and it did not suit their self-image to concede that civilization might already have been there.

During the 18th and 19th centuries, the newcomers first began to feel the prickings of guilt about the culture they'd destroyed . . . and, defensively, managed to persuade themselves that they hadn't really done it. Even despite the existence of journals by the early European explorers recording that many of these settlements were fully functioning at the time of the White Man's influx – and even much later, because some explorers were still encountering such settlements as late as the late 18th century – White Americans decided to believe it was impossible for the indigenes they'd nearly exterminated to have constructed such a civilization. The settlements and their associated artefacts must surely be the remnants of a now-vanished race. Hence arose the myth of the Moundbuilders, and the hunt was on to identify who those mysterious lost people might be and where they might have come from.

Plenty of spurious evidence was brought forward to support the various hypotheses. At the time it was thought that the Native Americans had arrived relatively recently across the Bering Strait from the Old World, and so it was thought a good idea to produce datings of the mounds that set their building back to before the Red Man's arrival. The Native Americans produced metal artefacts only in copper, whereas some of the artefacts associated with the mounds were done in other metals. Some artefacts were embellished with writing: not only was it "a well known fact" that the Native Americans had never developed writing, but in many of these instances there were characters that strongly resembled letters from Old World alphabets. And so on.

The scale of this exercise in self-deception is almost as impressive as the scale of the mounds. Just about every Old World culture you could think of was identified as the source of the immigration that had brought this great civilization to the Americas. According to Constantine Rafinesque (1783–1840) and Josiah Priest (1788–1851), among others, the mound-builders had been descendants of Noah, Priest specifying that they had been the offspring of Noah's son Shem. It hardly needs to be said that the Lost Tribes of Israel and refugees from Atlantis were popular additions to the list. Cotton Mather (1663–1728) was one who supported the Lost Tribes hypothesis, which may have originated in the book *De Extremo dei Judicio et Indiorum* (1567) by Johannes Fredericus Luminius.

A little sanity began to enter the debate with the publication in 1848 of *Ancient Monuments of the Mississippi Valley: Comprising the Results of Extensive Original Surveys and Explorations* by Ephraim G. Squier and Edwin H. Davis. Squier and Davis, unlike so many of their contemporaries, had thought it might be useful actually to investigate first-hand the sites that were the focus of so much frenzied speculation. In the conclusion of their exhaustive account, while they still maintained that the indigenous North Americans were far too primitive to have been responsible for such a magnificent outpouring of culture, they hazarded the guess that there might be some cultural link with the known civilizations of South America, such as the Aztecs and Maya.

Not until some decades later, when Cyrus Thomas (1825–1910) was commissioned by the Bureau of American Ethnology to try to solve the Moundbuilder "mystery" once and for all, was the false picture finally and forcefully shattered – and even then there was plenty of resistance to the unwelcome news. His conclusion in *Report on the Mound Explorations of the Bureau of Ethnology* (1894) was a statement of what is to us now the obvious: there was no reason whatsoever to think the Native Americans were not the Moundbuilders, and direct and overwhelming evidence that they were. Only a combination of wishful thinking, dogged self-delusion and frequent, near-transparent fraud had closed people's eyes to this.

But what about those artefacts inscribed with letters reminiscent of Old World alphabets? First, it must be borne in mind that it's not at all uncommon for occasional characters in unrelated alphabets to bear a chance resemblance to each other. Second, and far more important, almost all of the artefacts in question could quite easily be shown to be recent fakes, produced sometimes in a conscious effort to perpetuate the myth, sometimes as a means of fleecing the gullible, and sometimes just as practical jokes.

Yet the myth of the ancient Moundbuilders seems unable to die, and a version of it is still extant today in the efforts of those like the US marine biologist Barry Fell (1917–1994) who seek to "prove" that the ancient peoples of the Old World were frequent visitors to the New. In books such as *America BC* (1976), Fell claims evidence abounds to show that civilization was brought to the Americas by successive waves of Europeans, beginning about 3000 years ago. His most astonishing claim is perhaps that certain scored rocks could have been inscribed in Ogam; since Ogam was a written language that to the untutored eye looks much like random marking, it's hard to see how Fell could come to this conclusion. Quite why the White Man should *still* be so reluctant to accord the Native Americans their cultural due is a matter almost beyond comprehension.

The angel Raphael tells Adam and Eve of the Creation.
Done by Thomas Kirk, c1770.

CHAPTER 3

SURVIVAL OF THE BRIGHTEST

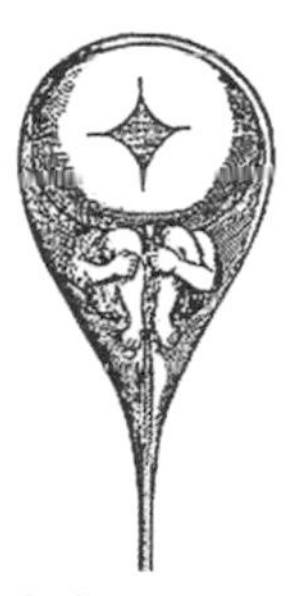

THE DEPRESSING RESULTS of a survey done in the UK at the beginning of 2006 by the BBC for its *Horizon* current-affairs series showed that, despite the best efforts of educators and the responsible media, and despite the near absolute unanimity of scientists on the subject of evolution by natural selection, still only 48% of UK adults think it is the correct explanation of the origin and development of life. If one takes the surprisingly high percentage of "don't knows" out of the picture, 54.5% of the respondents opted for Darwinian evolution, 25% for Creationism, and 19.3% for Intelligent Design (ID). (The percentages don't quite add up to 100 because of rounding.)

When asked which of the theories they'd like to see taught as part of the science curriculum (more than one choice permitted), 44% said Creationism, 41% said ID, and 69% Darwinian evolution. Although the final percentage is the largest of the three, it's still mind-bogglingly low, indicating presumably that the bulk of the Creationist and ID proponents wanted Darwinian evolution barred from the science classroom. There was also an age-related slew in the results, with people over 55 being significantly less inclined to accept Darwinian evolution.

Still, the UK results paint a brighter picture than the situation in the US. A 1999 Gallup poll there showed that only 18% believed in evolution as the sole explanation for the origin of the human species, while 38% believed exclusively in a Creationist explanation and 43% in some variant or other of Intelligent Design. (By way of comparison, according to a different poll, 68% of Americans believe in the existence of the Devil – about four times as many as accept evolution.) The poll also showed an exceptionally strong correlation between level of education and acceptance of the principle of evolution by natural selection: 65% of those with a postgraduate degree accept evolution. Turning that figure around, however, reveals a horrible fact: 35% of Americans with postgraduate degrees do *not* accept evolution. That there should be such a level of rank scientific ignorance even among the best-educated of US society indicates something extremely rotten at the heart of either the US educational system or the US media, or both.

Just to add to the general disarray of US school science teaching, in early 2006 Governor Jeb Bush (b1953) of Florida, brother of the President (who had himself, a few years earlier, caused consternation in the scientific and educational communities by advising the teaching of ID in schools in the name of "balance"), announced that he didn't want evolution to be part of the state's new school science standards. He didn't actually object to it being taught in schools: he just didn't think it should be a mandatory subject. What next? Maths classes don't have to mention pi?

The debate on the origins of Man is inextricably linked with ideas of evolution and of the age of the Earth. It is now generally thought that Man and the anthropoid apes evolved from common stock, the two strains separating several millions of years ago, quite possibly before either bore much resemblance to their modern forms. Even after the split, it seems likely that for some while the two species could, and did, interbreed. Certainly humankind's evolutionary history is complex.

That's the official story. The unofficial stories are many and diverse. Mythological/religious accounts of the origins of

Man are legion. The Kirghiz tribe believes we are the descendants of swine: not surprisingly, they have a taboo against eating pork. There are plenty of hypotheses produced in the developed nations that are equally as bizarre.

Despite the fossil evidence, could Man be a lot younger than the millions of years the palaeontologists talk about? This argument was easier to support in the days when the fossil evidence was scanty. In 1821 Cardinal Newman (1801–1890) wrote in his diary:

> Buckland [William Buckland, the eminent geologist] has just noted in his geological lectures the extraordinary fact, that, among all the hosts of animals which are found and are proved to have existed prior to 6000 years ago, *not one* is there which would be at all serviceable to man; *but* that directly you get within that period, horses, bulls, goats, deer, asses &c. are at once discovered. How strong a presumptive proof from the face of nature of what the Bible asserts to be the case.

There is a causal flaw in the argument, of course: if our ancestors had been around in the time of the dinosaurs, might they not have domesticated the local reptiles instead? Of course, if you believe some of the more sensationalist "archaeology" texts – not to mention some of the Creationist offerings – this is precisely what our ancestors did.

In similar vein, much more recently the Evolution Protest Movement has claimed it would have been impossible for human beings to have withstood the cold of an ice age; thus humankind can be no older than about 10,000 years. The Movement presumably disagrees with those authorities who maintain that, even in the most severe glacial period, large parts of the Earth's surface can enjoy temperate or even tropical conditions. And have the Movement's members never heard of Eskimos?

Maybe some of these ideas, too, should be taught to schoolchildren in the name of "balance".

However, before we step into the boggy marshlands of the discarded science connected with evolution, first we should take a look at some earlier concepts.

THE CHAIN OF BEING

This notion was born from the superimposition of religious beliefs on the results of philosophical musings and scientific researches. Essentially, the concept was this:

It could be seen that life existed at every conceivable level of simplicity and complexity, from Man at the top to "animalcules" at the bottom. Above Man were the angels and, finally, God; below the simplest living things was the *vis plastica* (see page 129) of the fossils. Thus each level of complexity could be envisaged as a link in a chain – and an essential link: if the link "cat" were missing, then no creature more complex than a cat could exist because the chain would be broken. Some links were obscure – e.g., those between simple organisms and the fossils (corals seemed a likely candidate) – but most could be established easily enough. The root of the concept goes back to the Aristotelian idea that, if living things were arranged in a hierarchy, there could be perceived a continuity between each lifeform and the next in that hierarchy, as well as to the Platonic idea that all possible types of lifeform exist.

The ramifications of the Chain of Being concept were tremendous, especially when the theory was taken in conjunction with that of Spontaneous Generation (see below) in the debate on the plurality of inhabited worlds (see page 197); since God was universal and since simple living things *must* occur on other worlds (because of Spontaneous Generation; besides, would God waste his creative energies on creating other worlds if not to populate them?), then so should more complex beings, all the way up to Man. But God had created Man in his own image. Thus other worlds must be populated by lifeforms identical with those of Earth.

The Chain of Being came into the limelight once more during the 17th and 18th centuries, when some of the heavyweights of European science lined up behind the concept, among them the German mathematician/philosopher Gottfried Leibniz (1646–1716) – famed as, along with Isaac Newton (1642–1727), the inventor of the calculus (although in fact the guts of the calculus had been worked out earlier by

Newton's tutor Isaac Barrow [1630–1677]) – the Swiss naturalist Charles Bonnet (1720–1793) and the French naturalist Georges-Louis Leclerc, Comte de Buffon (1707–1788). Perhaps its most influential adoption was by the Frenchman Jean-Baptiste Lamarck (1744–1829). Lamarck posited that not one but two Chains of Being existed, one for the animals and one for the plants. Unlike many of his contemporaries, in order to establish the position of a lifeform in one of these hierarchies he looked not at superficial resemblances but at functional resemblances, and in particular at similarities between the morphologies of the structures associated with those functions in different species. Since he could see smooth transitions in this context between one lifeform and the next, these studies led him on to infer that living creatures had evolved and indeed were constantly in a state of evolution. Although the mechanism he proposed for evolution, the inheritance of acquired characteristics (see page 133), was flawed, and although his evolutionary ideas were stifled by the powerful Georges Cuvier (1769–1832) – the idea of smooth, steady evolution was anathema to Cuvier's Catastrophist hypotheses (see page 45) – Lamarck still stands as an important precursor of Darwin and Wallace.

A very vaguely associated idea was the Law of Ancestral Inheritance (or of Blending Inheritance), which had it that the individual's make-up was attributable one half to the parents, one quarter to the grandparents, one eighth to the great-grandparents, etc., *ad infinitum*. Thus, according to the German botanist J.G. Koelreuter (1733–1806), when species produced hybrid variants, the character of the hybrid was an intermediate form of the characters of the two original species. That this was not in fact the case was observed by another German botanist, C.F. von Gaertner (1772–1850): although accepting that Koelreuter's ideal of blending characters was generally true, he saw that at least some hybrids displayed not an intermediate, compromise character but a mixture of the attributes of the two parent species. Charles Darwin (1809–1882) showed the general inapplicability of the blending hypothesis in his *Variation of Animals & Plants Under*

Domestication (1868), but it survived at least until the beginning of the 20th century.

The Balance of Nature

That the state of Nature was a delicate equilibrium sustained by the interdependence of all the plants and animals in it was a Victorian idea close to our modern concept of the ecosystem. The fallacy contained within the 19th-century notion is, however, readily apparent: the Nature it envisaged was static, whereas in reality Nature – i.e., any functioning ecosystem, however defined – is a dynamic entity, being in a perpetual state of flux. Thus the introduction of a new element – for example, the advent of an ice age – will not so much "upset the Balance of Nature" as force the creation of rafts of new ecosystems (alongside the destruction, clearly, of countless old ones) and thereby probably accelerate evolutionary change. Indeed, we could go further and say that the true Balance of Nature depends upon such periodic catastrophes, whether on a small or a global scale. A static Nature would in due course degenerate.

This is not, of course, to suggest that colossal and sudden catastrophes are to be encouraged: a full-scale nuclear war, for example, would destroy ecosystems on such a vast scale that recovery, if recovery were possible at all, would require millions of years.

Fossils

Once upon a time – and it was only about 250 years ago – there was considerable debate as to what fossils actually *were*. Some said they were unusual mineral formations, others that they were the remains of subterranean creatures, still others that they were the remains of organisms which had been destroyed in the Flood. The situation was chaotic.

In the 18th century Professor Johann Bartholomew Adam Johann Beringer (1667–1740) found apparent proof that fossils had been planted by God. And most extraordinary fossils they were: perfectly preserved lizards, frogs, scorpions . . . there was even a spider complete in its intact web, not to mention a stone bearing the Hebrew word for "God". Beringer wrote a book on the subject, *Lithographiae Wirceburgensis* . . . (1726). In fact, the "fossils" he found had been made and planted by a couple of his colleagues, J. Ignatz Roderick and Johann Georg von Eckhart (1664–1730), who had become irritated by Beringer's arrogance and pomposity. On hearing he was about to publish his book, they dropped heavy hints that he was the victim of a hoax, but he decided his informants were merely trying to prevent him achieving the glory he deserved, and dismissed the information. But we should be careful not to mock Beringer too readily, for in the early 18th century it was still reasonable to believe that the Earth had been created only a few thousand years earlier, and that there had thus not been enough time to allow for the processes of fossilization. Over a century later Philip Gosse (1810–1888) was still soberly presenting the notion that God had planted the fossils to kid human beings into thinking life on Earth had undergone a long evolutionary history. Beringer, then, was really a victim of not just his own arrogance but also the intellectual climate of his day.

Leaving Gosse aside, the picture was further clouded by the longstanding theory of Spontaneous Generation. For centuries it was widely believed that living creatures – and even more so fossils, lying as they did halfway between lifeforms and the inanimate – could emerge from nonliving materials by this process, which was also known as abiogenesis. The idea was based on observation. For example, if you left a barn full of grain for a couple of weeks, mice appeared – and from where else could they have come but from the grain, the air, or a combination of both? Likewise, if you left a piece of meat long enough, maggots appeared. It was only a small stretch to say that new species could emerge from nowhere in the same way.

Even today, many people would be prepared to countenance the spontaneous appearance of *small* creatures – e.g., maggots – although most would (one hopes) be less enthusiastic about Spontaneous Generation than some past theorists.

When Aristotle (384–322BC) classified the animal kingdom, partly according to the mode of reproduction, he described as abiogenetic such organisms as eels, sponges and certain fish. Francis Bacon (1561–1626) touched on the abiogenesis of thistles from earth – odd, because thistledown is surely fairly obvious. Similar myopia was shown by Robert Hooke (1635–1703), who said mushrooms and moulds must be abiogenetic, since he could see no seeds for them through his microscope; modern microscopists complain of the difficulty of examining these organisms because of all the blasted spores getting in the way!

Francesco Redi (1626–1697) first tackled the maggots problem. He showed that they developed in rotting meat only if flies had been in contact with it: no flies, no maggots. But he still thought it likely that abiogenesis occurred in other instances.

William Harvey (1578–1657), the discoverer of the circulation of the blood, had been an early opponent of abiogenesis, holding that all living things were born from eggs; and the coming of the microscope proved him right, in most cases. But this new instrument permitted also a revival of the theory in a new guise. Anton van Leeuwenhoek (1632–1723) first observed in 1674 what he called "animalcules" – protozoa. Here were tiny organisms which seemed to generate spontaneously.

Obviously, Redi's experiments should be repeated. They were . . . and they seemed favourable to abiogenesis. John Needham (1713–1781) published in 1745 *An Account of Some New Microscopical Discoveries*, in which he told how gravies heated in sealed tubes for lengthy periods (thus presumably killing any organisms present) were later able to "give birth" to organisms. He and the Comte de Buffon came up with the notion of "vital atoms". Living things consist of a mixture of vital atoms and nonliving material; at death, the vital atoms return to the ecology, eventually to become part of other creatures. But the Italian biologist Lazzaro Spallanzani (1729–1799) disagreed strongly. He repeated Needham's experiments, but heated the gravies more vigorously and for longer, and sealed his containers more carefully. He found the mixtures did not become infected with microorganisms. His opponents countered that,

Fossils captured the popular imagination during the 19th century.

in heating the gravies, he had also heated the air above them, thereby destroying some as yet unidentified "vital principle" in it and so invalidating his results. This was quite true, in a way, but of course no one knew about airborne microorganisms in the 18th century. The matter was still unresolved as the 19th century dawned.

In his argument with Spallanzani, Needham had used a well established scientific technique to help his case: he had fiddled his experiments. He did this by simply ignoring any suggestion that he might not be sealing his containers properly. There's a suspicion that he *did* repeat the experiments with properly sealed flasks, but kept quiet about the results: the natural response to the accusation that he hadn't done his experiments right the first time would have been to repeat them to show that he had. Spallanzani certainly responded thus after Needham and Buffon claimed he was destroying the "vital force" by boiling his broths too long: he countered the suggestion by performing exactly the series of experiments which Needham should have done (or perhaps did) and, in so doing, found that some microorganisms could survive brief boiling, and others without oxygen – two profound discoveries largely ignored for several decades.

Louis Pasteur (1822–1895) researched fermentation during the 1860s and noted that, where fermentation failed, it was due either to a lack of the necessary microorganisms or to circumstances which made it impossible for them to develop normally. Where did these organisms come from? Either they were omnipresent in the atmosphere or they generated spontaneously, and he set out to discover which. (As early as 1762 Viennese physician Marcus Plenciz proposed that the air is aswarm with invisible animalcules, which cause disease. This bullseye, alas, accompanied his suggestion that they give rise also to gnats, beetles and leeches.) First of all Pasteur showed that exposing samples to air gave rise to hordes of microorganisms. He then placed sterilized samples in containers to which the air had access, but only through a long narrow tube, bent like a flush-toilet's U-tube. Any airborne microorganisms would collect at the bottom of the "U". The result – no putrefaction.

The final blow came around 1880 when John Tyndall (1820–1893) – the Irish physicist who explained why the sky is blue – devised an apparatus which allowed him to tell whether or not air was pure. Sure enough, he found that organic matter does not putrefy in pure air.

The biggest riddle presented by fossils was, obviously, that they looked like organic remains but seemed to be made of rock. An early explanation of the phenomenon was put forward by Aristotle, who attributed seismic activity to the workings of subterranean winds, and by extension proposed that the Earth's metallic resources were produced by moist gusts from beneath the surface, nonmetallic minerals and fossils by dry ones. To explain the organic appearance of fossils he was forced to invoke a "moulding force", *vis plastica*, at work within the Earth, imitating the activities of Nature on the surface

There were several reasons why the rejection of the organic origins of fossils, and the acceptance of the *vis plastica* theory in some form or another, survived until the 18th and even the 19th century. One was that fossils are often found at depth within rocks: how could they have got there? If organic, they certainly could not have been put there during the Creation, for *Genesis* insisted that God made the dry land before he created life. In the Middle Ages it was suggested that fossils were the remains of plants and animals destroyed during the Flood, but it was generally agreed that the Flood had not lasted long enough to be solely responsible. Again, fossils were often markedly different from living species, which might imply that some species had become extinct. But this would be to imply that God had created "too many" species in the first place – had overstocked the larder, as it were. As God didn't make mistakes, the conclusion was that fossils could not be of organic origin. (In fact, opponents of the *vis plastica* idea got around this by pointing out that there might be living examples of fossil species in unexplored parts of the world.) As always, matters were not helped by the belief that the Earth was only a few thousand years old.

In 1663 Robert Hooke applied the microscope to the problem. He compared petrified wood with ordinary wood and found that, while there were certainly differences between the two, the structural similarities were such that a common origin had to be assumed. In fact, his account of the processes of fossilization was remarkably close to modern ideas. However, most of his contemporaries remained sceptical.

The *vis plastica* idea gave rise to some unusual offshoots, such as the notion of vegetable gold. To the ancients, since deposits of metals were found underground, it seemed reasonable that they grew there – like potatoes, for example. In the Middle Ages the notion was developed further: precious metals, gold in particular, dwelled underground in malleable form. The roots of plants grew through these deposits, and some plants veritably sought them out, drawing the gold up through their roots and incorporating it into their leaves and branches. Foremost among the gold-hungry plants were grapevines.

The notion was bolstered by occasional genuine discoveries by viticulturists of gold filaments and sometimes larger pieces among their vines. The roots of plants can bring items to the surface, and vine roots are especially good at this. Centuries earlier, their lands tormented by wars, people had buried gold artifacts for supposed safekeeping; war being war, often enough these hoards were not recovered afterwards, and remained undiscovered until the vine roots unearthed some of the smaller pieces.

Well into the 18th century the idea of vegetable gold persisted, and grapes containing fragments of gold were much prized among the affluent. Many of these grapes were, of course, simply fakes: ours is not the only age in which affluence and gullibility go hand in hand. In other instances, the flecks of gold weren't gold at all but the yellow eggs of certain insects.

When saffron started to be available in the West, one of its alternative names became – and remains – vegetable gold. This was presumably in reference not just to the colour but, more particularly, to the spice's value.

A great puzzle with the fossil record was that there seemed to be an evolutionary progression from the fossils found in the oldest rocks to those found in the youngest. Evolution wasn't considered because, once again, the Earth didn't seem to be old enough to permit it. The French anatomist Georges Cuvier suggested a compromise between evolutionary ideas and Creationist ones: he proposed that there had been a *series* of separate creations, each following some cataclysmic extinction of earlier lifeforms. In this way he thought the changing nature of fossils could be explained. It was rather a desperate compromise.

But the finest elaboration of *vis plastica* ideas came in 1857, with the publication of Philip Gosse's *Omphalos*. Gosse believed the Earth had been created in 4004BC, but he accepted that geologists, astronomers and palaeontologists were producing evidence which seemed to show that certain events had taken place long, long before this. Moreover, there was the matter of Adam's navel, about which great debate had raged (*omphalos* means "navel"): did Adam have one or not? Gosse suggested that he did, and that he had been born fully grown, with such adult characteristics as hair, normally assumed to be the result of prior growth.

The paradox could easily be resolved, Gosse decided. Learning nothing from the tribulations of the unfortunate Beringer, he proposed that God had created the world, the Universe and Adam *as if* they had had a previous history! All that the astronomers and geologists were showing was the completeness of God's artifice as they detected more and more details of His conscientious craftsmanship. None of these details implied that events had actually occurred before the Creation, in 4004BC. God had planted the fossils in the ground as a part of His scheme. The delight about this theory is that it can never be either proved or disproved. The Creation could have occurred only five minutes ago, and no one would be any the wiser.

THE EVOLUTION OF EVOLUTION

On the third day God made the seed-bearing plants and the fruit-bearing trees, on the fifth the marine animals and the birds, and on the sixth wild creatures, reptiles and cattle, and finally Man, in God's own image. Thus according to *Genesis*. It would seem the author of *Genesis* had some vague idea of evolution, of the idea that complex organisms arrived on the scene after simpler ones. If that author was indeed Moses, then here we have evolutionary concepts current in the 13th century

BC. And Lucretius (*c*99–55BC) wrote: "In those days many species must have died out altogether and failed to reproduce their kind. Every species that you now see drawing the breath of life has been protected and preserved from the beginning of the world either by cunning or by prowess or by speed . . ." It would seem Lucretius could stomach the idea of extinctions, and was thus partway along the path towards an evolutionary theory.

Other early ideas were less plausible. Anaximander (*c*610-*c*545BC) thought the first human beings were born, fully grown, from the wombs of fish. Empedocles (*c*490-430BC) is reported to have thought the earliest creatures were not wholly formed, but consisted of unconnected limbs. His notion would be the object of much mirth for many centuries (imagine a hand flapping along the primaeval beach looking for an arm to join onto) until the proponents of ID came up with something very similar nearly 2500 years later.

The Comte de Buffon argued that the Universe had been created by God but that God had then left it on its own to evolve according to the natural laws He had created. Mention of Buffon's idea raises an important question: since evolutionary ideas have been around for millennia, why were 19th-century theorists such as Charles Darwin (1809–1882) treated as heretics? There is no essential conflict between the idea of the Creation and the theory of evolution by natural selection. It was widely felt that the Creation had been a fairly swift affair, because God, being timeless, need take no time over it. But by the same argument a mere four and a half billion years is no time at all to God. Which would He prefer – a single act of instant Creation, or a mixing of the various ingredients, leaving time to do the rest? Attempting to answer such questions comes dangerously close to trying to read God's mind! But that is exactly what the devout Creationists were doing – as many indeed still are.

The irrelevance of the Fundamentalist argument against "evolutionary Creation" is exemplified in these words from Pope Pius XII (1876–1958):

> These figures [for the age of the Earth] may be surprising but they do not contain any different concept, even for the simplest believer, from that in the first words of *Genesis*: "In the beginning", which means the beginning of things in time. The figures of the scientists give to these words of scripture a concrete and at the same time a mathematical expression . . .
>
> Anyone examining these problems seriously from the point of view of modern scientific knowledge, must give up the idea of wholly independent and autochthonous material, uncreated or self-created, and must reach the conception of a creative Mind. With the same clear and critical eye, with which he judges facts, he will recognize the work of a creative omnipotence.

In 1809 and 1816, long before Darwin's and Wallace's public presentation of the theory of evolution by natural selection (in 1858), the French naturalist Jean Baptiste de Lamarck put forward a different evolutionary theory. Where Darwinism relied upon natural selection as a mechanism (random mutations occur in populations; individuals with favourable mutations are more likely to survive to breed, thereby possibly perpetuating the mutation), Lamarckism relied upon the hypothetical ability of individuals to inherit characteristics that had been acquired by their parents; to turn this round, the notion depended on the speculation that changes which had occurred in an individual during its lifetime could be transmitted to its offspring.

Lamarck's four evolutionary laws were:

- a characteristic of "life" is that it tries to increase the size of the body that it occupies, as well as the individual parts of that body, until a limit is reached;
- new organs or members of the body are produced as a response to a new-felt need;
- the size and strength that organs develop are in proportion to the frequency and vigour of their use;
- acquired characteristics can be transmitted to the next generation.

These can be summed up in simplified form by saying that behavioural changes cause physiological changes, and these physiological changes then become permanent parts of the evolutionary heritage.

It would be a mistake to think that Lamarckism sprang fully fledged from Lamarck's fertile brain: similar ideas can be traced back as far as *Genesis*. In the 16th century Gerolamo Cardano (1501–1576) believed that his fondness for river crabs came about as a result of his mother eating them while she was pregnant with him. In the same century, Michel de Montaigne (1533–1592) came up with a poser for evolutionists. He had inherited from his father a knack for producing kidney stones . . . but his father did not start to suffer from kidney stones until Montaigne was aged 25, and Montaigne himself did not start to produce them until he was 45. What mechanism, then, could explain his inheritance of his father's characteristic? Much later, in the decades leading up to Lamarck's formulation, both the Comte de Buffon and Erasmus Darwin (1731–1802), grandfather of Charles, put forward evolutionary theories that were much the same as Lamarck's. Lamarckism was influential enough that Charles Darwin himself, while maintaining that natural selection was the prime mechanism of evolution, at least for a while invoked Lamarckism as a subsidiary mechanism, a means of evolutionary fine tuning, as it were. And Darwinism did not replace Lamarckism overnight: the two theories were rivals well into the 20th century, and even today the debate cannot be considered entirely over.

Among the distinguished supporters of Lamarckism was the US palaeontologist Edward Drinker Cope (1840–1897), who bravely worked to popularize evolutionary ideas in a land where evolution was still, to the mentally inflexible, dangerously close to heresy if not outright blasphemous. One of Cope's special interests was the mind, and so it was natural for him to assume that evolutionary variations could be effected by the mere force of willpower. This notion opens up the delightful prospect of humanity having the ability simply to *will* itself a brighter evolutionary future, or certainly of individuals being able, by a conscious exercise of mind, to improve all sorts of characteristics of their offspring.

A better known case than that of Cope concerns the Austrian biologist Paul Kammerer (1880–1926). Some of his experiments seemed to prove that acquired characteristics could indeed be inherited, and on quite a grand scale; to his own apparent embarrassment he was publicly acclaimed as the successor to Darwin. But in 1926 it was shown that one of his most influential experiments – concerning a seemingly inherited mark on the hand of a midwife toad – had been fiddled, and this obviously cast doubt on all the rest of his results. Arthur Koestler (1905–1983), in his book on the whole imbroglio, *The Case of the Midwife Toad* (1971), argued persuasively that it was not Kammerer himself but one of his assistants who had "helped" the experiment. Whatever the truth of the matter, Kammerer committed suicide in the same year as the exposure.

In the Western world, that more or less spelt the end of Lamarckism as a scientifically respectable theory – although some much more recent researches seem to show that, at least at the genetic level, acquired characteristics may indeed be transmitted. In the Soviet Union, however, Lamarckian ideas held sway for much longer. There Trofim D. Lysenko (1898–1976) adopted wholesale the Lamarckian ideas of the horticulturalist Ivan Michurin (1855–1935) and gained Stalin's favour. Soviet biologists who disagreed with Lysenko's line could do so in either silence or Siberia. It was for this reason that, until Lysenko's fall in 1964, Russian genetics lagged by decades behind that of the West.

Consideration of the inheritance of acquired characteristics was confused by a set of experiments reported in the 1950s. Although many believe these experiments were refuted, the truth is that the matter was simply allowed to slide into abeyance; the meaning of the results remains a mystery.

The experiments were done by US psychologists James McConnell and Robert Thompson using flatworms (planaria). Planarians are among the simplest vertebrates, and have the

A much later portrayal of the medieval worldview: beyond the edge of the physical U

the realm of God. From *Weltall und Menschheit*, edited by Hans Kraemer, 1907.

distinction of being the simplest known creatures to possess a brain – well, sort of: there is a little cluster of cells near the head of the 2cm-long worm that could be called a brain in one's more charitable moments. McConnell and Thompson set up an experiment whereby the flashing of a light was coupled with the passage of an electric shock through the water in which individual planarians swam. The shock caused the planarians to curl up in a distinctive fashion. After this had been done a number of times, the planarians – in a manner reminiscent of Pavlov's dogs' reaction to a ringing bell – would curl up in response to the light alone, without the electricity being administered. In other words, even with a brain so rudimentary that many people wouldn't describe it as a brain at all, the planarians were capable of *learning*.

This result was startling enough, but over the next few years the two researchers went further. The little worms have the capacity, when cut in half, of regrowing as two separate, living worms. It was plausible that the worm regenerated from the head/"brain" end of the worm would retain the memory of the light and its associated shock, but the same could surely not be true of the regenerated tail end, could it? The astonishing result reported by McConnell in 1959 was that the regenerated tail ends showed just as good – and high – retention of the memory as did the head ends. This suggested that the planarians had some means of storing memories other than in the brain, or that a planarian regenerated from a tail end was an exact replica of the original, including the brain hardwiring associated with memory.

McConnell thought there must be some sort of biochemical explanation: that the memories were stored as a chemical throughout the planarian body. In order to test this hypothesis, he and his team trained a batch of planarians with the light and the shock, then cut the unfortunate individuals up and fed them to other, untrained planarians. (This is reminiscent of the hypothesis that you can gain a person's wisdom and other qualities by eating their brain – see page 162.) By way of control, the team fed further planarians with gobbets of *un*trained planarians.

This time the results were really a shock. The planarians fed on trained individuals showed, when tested, a significantly higher rate of curling up in response to the flashing light than did the planarians in the control group. The effect wore off after a few days, but it was quite distinct while it lasted.

The publication of these results in 1964 was met with widespread incredulity. Almost immediately, a couple of teams – one led by Melvin Calvin (1911–1997), who was awarded the Nobel Chemistry Prize in 1961 for his work on photosynthesis – repeated the experiments, with results that were at best confusing, and in 1966 the journal *Science* published a letter from 23 distinguished researchers saying they could find no evidence of memory being transmitted chemically, as McConnell suggested was the case. That was generally taken to be the end of the matter . . . except that, of course, the letter missed the point. Assuming McConnell wasn't faking his results – and no one has ever suggested he was – then, while he might have been wrong about the mechanism of memory transfer being of chemical nature, this does nothing to alter the fact that the memory transfer apparently occurred. The whole affair was rather left in limbo, and no proper explanation has ever been put forward.*

Popular and, to a certain extent, scientific belief that there "might be something in" Lamarckism continues to this day. And in the field of machine evolution – whether talking in terms of software or, with self-replicating machines, hardware – it is clear that Lamarckian-style mechanisms must be very important, and indeed overwhelmingly so.

A full account of the great 19th-century debate over Darwinism is beyond the scope of this book, but a few of the arguments put forward in the debate deserve some mention here. If indeed there *was* a debate! Thomas Bell (1792–1880), President of the

* In 1985 McConnell was a target of the Unabomber, Ted Kaczynski. Although suffering only minor injuries, he remained partially deaf for the last few years of his life; he died in 1990.

Linnaean Society, noted at the end of 1858, the year in which Darwin and Wallace had presented the theory to the Society, that the year had not "been marked by any of those striking discoveries which at once revolutionize, so to speak, the department of science in which they occur". The Fellows apparently shook their head in wonder that their President could be so oblivious to what was going on.

But to think of 1858 as the year in which the debate started would be misleading: as we have noted, evolutionary ideas were current long before that, and were certainly coming to some kind of a focus around the start of the 19th century. So were rebuttals. Georges Cuvier was one who proved to his own satisfaction that evolution did not and had not happened. He pointed out that every species is extremely well designed for its niche in life; even a minor change in that design would be fatal for the species concerned. He therefore envisaged a separate Creation not only for each species but also for each *organ* of each species – again, shades of Empedocles, and certainly a precursor of ID. Cuvier's was an astonishing inference in the light of the fact that his own, original, pioneering researches on fossils were showing that, the deeper the stratum in which a fossil was found, the less like modern lifeforms the fossil creature generally was. Cuvier's only excuse must be that he assumed the youthfulness of the Earth.

This reluctance to accept evolutionary ideas, whatever the evidence, was shown by the Estonian natural historian Karl Ernst von Baer (1792–1876). He independently put forward an evolutionary theory – in 1859, the year *Origin of Species* was published. But, while von Baer could conceive that similar animals might have evolved from common stock – he said the similarities indicated "kinship" – he violently opposed Darwin's suggestion that *all* lifeforms, humankind included, might have evolved from a common ancestral form. There were many, many others who shared his slightly illogical views, his half-acceptance of evolution.

A more intriguing counterargument was put forward by the UK physicist Baron Kelvin (1824–1907). He was interested in the rate of the Earth's cooling from its original molten state; and pointed out that, on the basis of thermodynamics, the Earth could not be more than 100 million years old, Moreover, conditions even a mere million years ago must have been very different from those obtaining now: the Sun would have been far hotter then and so, of course, would have been the Earth's surface. Naturally these different conditions must have produced a set of lifeforms quite distinct from any we know today; and so to talk of the possibilities of evolution from that population to the modern one was senseless. Of course, Kelvin could not know that radioactive elements within the Earth have the effect of keeping our planet warm: his calculation of the Earth's age was out by a factor of about 46.

The UK palaeontologist Sir Richard Owen (1804–1892), who ran up a remarkable total of errors, misconceptions and wrong conclusions during his prominent career, seems to have opposed Darwin primarily for reasons of pride rather than of intellectual conviction: he realized that Darwin's reputation might eclipse his own. From 1856 he was Superintendent of Natural History at the British Museum, and from this power base he did much to hamper the course of UK palaeontology and evolutionary studies for decades.

One of Owen's key hypotheses was that of the archetype. He had noticed that often a similar anatomical structure will be present in a number of different vertebrate species, but with a different function: an arm and a wing and a flipper are all strongly analogous, but they do different things. (This fact is, incidentally, a rebuff to the notion presented by ID advocates that structures must appear for a specific purpose. Clearly structures can appear for one purpose and be adapted during evolution to quite another.) On the basis of his observation, Owen speculated that there must be a basic template for the vertebrates that God had created and then proceeded to modify when forming the diverse members of the vertebrate group. This template Owen called the archetype. He was never in fact able to offer any evidence for the existence of the archetype, nor even to spell out the implications of the concept, but to him it offered proof that the Divine hand was responsible for living creatures, not this newfangled evolution everyone was talking about.

Another piece of anti-evolutionary evidence he offered, on the basis of his dissection of a gorilla, was that apes lacked an important brain structure that humans possessed: the hippocampus minor. Besides, apes were evidently four-footed animals, unlike the bipedal human. The staunch evolutionist T.H. Huxley (1825–1895) became suspicious of Owen's claims, and was eventually able to perform some ape dissections himself. He reported triumphantly that the ape brain does indeed possess a hippocampus minor; furthermore, his anatomical studies showed that Owen was wrong, too, about the quadrupedalism of the apes: their forelimbs ended in hands, not feet as Owen had asserted.

Among further Owen hypotheses that bit the dust rather rapidly was his notion that the head is really just an extension of the backbone – i.e., that the skull is merely a modified vertebra.

But there are still some of us who are uncomfortable about the idea that our ancestors might have been creepy-crawlies (that we "descended from the apes" is a misleading simplification). Perhaps we need to take reassurance from the remark by J.M. Tyler that: "Even if we are descended from worms, they were glorious worms."

Muddying the waters of evolutionary discussions was the notion of ontogeny recapitulating phylogeny. That the alterations in form of the developing embryo might mirror the evolutionary history of the species is an idea directly traceable to the Biogenetic Law enunciated by 19th-century German natural philosopher Ernst Heinrich Haeckel (1834–1919), and is in many ways not indefensible. After all, in its early stages, the human foetus has structures called gill slits (these are non-functional, and not to be confused with gills proper), which would seem to echo our distant marine ancestors; and the likeness of all vertebrate embryos in their early stages might seem to reflect the fact that all vertebrates share common or similar ancestors if one traces their evolutionary roots back far enough.

Throughout the latter part of the 19th century and well into the 20th, Haeckel's doctrine that "ontogeny recapitulates phylogeny" did not completely hold sway but was at least treated with a certain respect. Some of his claims were overtly over-enthusiastic: having observed the gill slits, he believed the embryo went through a completely fishlike stage during its development. He had based his ideas in part on those of Karl von Baer, the Estonian founder of comparative embryology. Von Baer it was who had done detailed work on those very similarities between young vertebrate embryos; but von Baer steadfastly refused to accept Darwinian evolution. To Haeckel, ontogeny was a glorious proof of Darwinism.*

The mechanisms for the recapitulation were two: condensation and terminal addition. "Condensation" referred to the way in which the early ancestral forms were pushed closer and closer to the beginning of the foetus's development as the species evolved further. "Terminal addition" described the way in which evolutionary change came about as further developmental stages were tagged on to ancestral ontogeny: a human embryo, for example, would go all the way through fish, reptile, mammal, primate and hominid stages before the last, distinctively human touches were added.

One of the more extreme extensions of Haeckel's ideas was the Maturation Theory proposed by the US psychologist and educationist Granville Stanley Hall (1844–1924), who applied the recapitulation notion to the development of growing children, who, he claimed, passed through stages directly reflecting the stages of human evolutionary history. However, although one is often tempted to describe one's children as lit-

* Darwin himself seems to have been less impressed by Haeckel's theories: although in the third edition [1861] of *On the Origin of Species* he paid brief tribute to Haeckel's work on phylogeny in general, he never took on board anything about the Biogenetic Law. And, although Darwin did make use of the anti-evolutionist von Baer's results, he reached different conclusions from them – to von Baer's fury.

tle reptiles, the idea that this is literally true does not hold up to even cursory examination.

Occasional attempts are made to rescue Haeckel's Biogenetic Law from the litterbin of history, their grounds being that there probably *is* something in the theory and that the baby has been cast out with the bathwater. But such attempts are few and far between. That said, study of the developing embryo can be of considerable importance in evolutionary studies. The similarities shown by the developing embryos of different species can tell much about the evolutionary relationships between those species that would not necessarily be immediately apparent from study of their adult forms.

The Biogenetic Law was brought back into the limelight in 2000 with the publication of the book *Icons of Evolution: Science or Myth? Why Much of What We Teach About Evolution is Wrong* by Jonathan Wells, a Unification Church (Moonie) member who has stated that he studied for and attained his qualifications as a biologist with the specific aim of assailing Darwinism. In his book he seems to claim that modern embryologists and evolutionists still adhere to Haeckel's Biogenetic Law and indeed are conspiratorially protective of it in their publications, biology textbooks included, and he attacks them for their dishonesty on that basis. Unfortunately for his argument, the Biogenetic Law has been universally rejected by embryology for a century or more, so he is attacking embryologists for a view they do not hold. The textbooks he criticizes for supporting the Biogenetic Law, through either illustration or text, in fact do not do so: where it is mentioned at all it is in a strictly historical context, as it should be. His claim that Darwin based his embryological arguments in *On the Origin of Species* on Haeckel's theory is typical of Wells's technique: Haeckel's theory was not published until 1866, in *Generelle Morphologie*.

Further confusing the picture was the concept of telegony. According to this theory, the offspring of a female may be affected not only by their sire but also by previous males with whom she has mated. The theory may have given rise to the widespread taboo on a widow marrying her dead husband's brother: it might be biologically unwise in that future offspring would be, as it were, incestuous. Conversely, the notion may have given rise to the opposite – and again widespread – practice whereby widows should indeed be married by their dead husband's brothers: any resulting offspring would be as close as possible to the further children he might have sired had he lived, inheriting a mixture of his own attributes with those, presumably similar, of his brother (and the woman's, of course).

The notion of telegony dates back at least to Aristotle. Surprisingly, though, it has survived into the present century even in cultures that would normally regard themselves as scientifically sophisticated. For example, some dog-breeding boards, defying considerations of even the most elementary genetics, exclude from pedigree consideration the offspring of bitches that have earlier produced litters sired by dogs of the "wrong" breed. The reason for this survival may be that, while the notion reveals itself to be nonsensical after only a moment's thought, it is superficially plausible enough that it is very rarely *given* that moment's thought. Even Darwin fell into the trap: he debated in all seriousness a *cause célèbre* of the time whereby it was claimed that a mare had been mated with a quagga (a now extinct relative of the zebra) and later, on being mated with an Arab stallion, had produced striped offspring. (In the event, the Scientific Method triumphed: the experiment was repeated, and the results were of course negative.)

Dinosaur fossils in Colorado. From *The Graphic*, 1871.

In the light of telegony, it was of course important for a man to ensure that the woman he married was a virgin; otherwise his heirs might not truly be of his blood. This explains the otherwise extremely curious yet frequent insistence on the part of kings and lords that their spouses should not be widows or divorcees.

Among others who had evolutionary ideas before Darwin was the German naturalist Hermann Schäffhausen (1816–1893), author of the 1853 paper "On the Stability and Transformation of Species". Schäffhausen was involved in the initial furore over Neanderthal Man, remains of whom were discovered for the first time in a lime quarry near Dusseldorf in 1856. A local schoolteacher, Johann Karl Fuhlrott (1804–1877), was the first to realize the significance of what had been found, and he took the evidence to Schäffhausen, who wholeheartedly agreed with him that this was a major archaeological discovery. In 1857 the two men presented their case to an audience of distinguished scientists in Kassel at a meeting of the Natural History Society of Prussian Rhineland and Westphalia. Carefully Schäffhausen outlined his reasons for believing that the bones represented a human being from far earlier in antiquity than human beings had been thought to exist; he went further, and portrayed this long-lost race of humans as short and brutish.

The distinguished scientists flatly disbelieved him, insisting that the bones must be modern. One of those scientists was very distinguished indeed: Rudolf Virchow (1821–1902), whose pioneering work on cell pathology was a major contribution to medicine and who was the first physician to document leukemia and embolism. Virchow was a convinced polygenist (see page 150), and declined to believe in the sort of transmutations that just a few years later would be recognized as essential to the process of evolution by natural selection. The notion of a prehistoric form of human physically different from the modern version was thus anathema to him. His judgement on the remains was that they were relatively recent, but that the unfortunate individual had suffered from various gross disabilities that had deformed his bones. (Virchow was half-right: the man whose remains had been found *had* suffered disabilities. It was just that he had done so thousands of years earlier than Virchow was prepared to admit.)

With Virchow in steadfast opposition to any conceit that Neanderthal Man might be ancient, the whole affair was in danger of being entirely forgotten. A few years later, however, in 1861, the UK anatomist George Busk (1807–1886), realizing the importance of the paper Schäffhausen had published about the discovery, translated it for the *Natural History Review*. Not only was Neanderthal Man brought back into the spotlight, his antiquity accepted, but naturalists began to reconsider other unusual skulls that had been found earlier – one in Belgium in 1830 and another in Gibraltar in 1848. Finally T.H. Huxley had the chance to investigate the matter, and he confidently pronounced that Neanderthal Man was a human ancestor. Still the argument raged on, with evolutionists accepting Huxley's judgement, moderate anti-evolutionists and the undecided maintaining that Neanderthal Man was indeed ancient but unrelated to modern humans, and the solid anti-evolutionists stubbornly insisting that Virchow's analysis was correct.

Embarrassingly for the latter, Neanderthal specimens began to turn up with increasing frequency elsewhere in Europe; in reality, of course, they had been turning up all the while but people had just assumed they were the bones of old burials, if it was even realized these were human bones at all. (As Brian Regal suggests in *Human Evolution: A Guide to the Debates* [2004], it might be an idea for someone to check out the profusion of supposed saintly relics in European churches.)

The image of the Neanderthal as a brutish, stooped, shambling, muscular, moronic, apeman savage owes much to the interpretation by French archaeologist Marcellin Boule (1861–1942) of an important 1908 find of remains in a cave called La Chapelle-aux-Saints. The image remained prevalent for a very considerable while – it's still sometimes to be found in paintings – primarily because of Boule's position of influence at Paris's Musée Nationale d'Histoire Naturelle, and the politics he played there, successfully, to establish a position of personal dominance over French anthropology. His vanity served him ill when later in 1908 the Swiss archaeology hobbyist Otto Hauser (1874–1932) unearthed in the Dordogne the first remains from the Mousterian culture (a somewhat later Neanderthal culture than the Acheulian one): because Hauser was an amateur, Boule couldn't be bothered to accept an invitation to go along and supervise the final excavation, and so missed out on a major find. Peeved because only a few German anthropologists turned up for the event – no Frenchmen at all – Hauser sold the remains to them, losing them for France.

Similarly a fantasy is the popular image of the Neanderthal carrying a great knobbly club. Presumably the Neanderthals and Cro-Magnons did use clubs from time to time, but there is no extant evidence of their ever having done so.

The discovery of Cro-Magnon remains in 1868 in France complicated pictures of human descent, for these skeletons seemed to be at least as old as the Neanderthals yet were barely distinguishable from modern ones. Many archaeologists – Boule was one of the exceptions – had formulated a simple, linear model of human evolution, from the slouched Neanderthal to the gloriously upright European, with all too often the "lesser races" (meaning anyone coloured) being regarded as some kind of intermediate stage between the two. The matter came to the fore in 1901, with the discovery of a further Cro-Magnon site, this time in Italy. It seemed impossible to fit the Neanderthals into the linear model, and they were identified as a branch of humanity that had led to an evolutionary dead end. Boule and others foolishly extended this notion, though, speculating that there must have been an ancestral human stock, the "presapiens", that was virtually identical to the Cro-Magnons and modern humankind yet predated the Neanderthals. The presapiens had evolved rather suddenly from . . . well, here the concept broke down a little.

An evolutionary hypothesis somewhat out of the mainstream was presented by the eminent US zoologist Henry Fairfield

Osborn (1857–1935). He speculated that, when a new type of lifeform first appeared on the scene – by some sort of Spontaneous Generation – it had completely generalist characteristics; it also had a mysterious property which he called "race plasm". Only as the lifeform moved away during the succeeding generations from its locality of origin did different subgroups begin to develop more specific attributes, adapted to the various environments in which they found themselves. The effectiveness with which the subgroups could adapt in this way was governed by the quality of the original race plasm. However, the more specialized the adaptations of the members of a subgroup to cope with a particular environment, the less able that subgroup would be to adapt further should the environment change. Thus, once a subgroup had become especially adapted to that environment, it was unlikely to migrate any further away than this from the group's locality of origin. You could thus trace patterns of (say) an animal group's evolution by assessing the degree of adaptive specialization in its various subgroups from around the world and essentially plotting your results on a map, backtracking until you found the species' point of origin. It seems never to have occurred to Osborn that his judgements of degrees of specialization were entirely subjective, so that all of his efforts were entirely valueless. Certainly, using his system, he was able to demonstrate to his own satisfaction that the mammals had originated in Asia, which – surprise, surprise – was what he had believed in the first place.

When it came to one particular form of mammal, human beings, Osborn's theorizing had to be further engineered to take account of the fact that he believed Nordics were the best and purest and most "generalized" of all the human breeds. This created a big difficulty. If all humans had descended from Cro-Magnons and/or Neanderthals, then those earlier forms would by definition have had to be more "generalized" and overall *better* than his ideal Nordics. And if humanity had evolved from an ape-like ancestor the picture grew even darker. Drawing upon not just pseudoscience but theology for his inspiration, Osborn thus created the concept of the "Dawn

Man", an idealized and entirely hypothetical original human species from which the Nordics were descended. All of the other races, being but an inferior imitation, were descended from different original stock, which for all Osborn cared might be ape-like. Essentially he was, for the purpose of reinforcing his own racist preconceptions, reinventing the discredited hypothesis of polygeny – that Man's different races had different origins – although he called his own model "orthogenesis".

PANGENESIS

Charles Darwin produced several false theories both before and after the publication of his theory of evolution by natural selection; best known of these is his pangenetic hypothesis. This was his attempt, contained in an appendix to *The Variation of Animals and Plants Under Domestication* (1868), to explain the Lamarckian phenomenon of the inheritance of acquired characteristics, a phenomenon he at the time accepted.

Darwin proposed that bodily cells constantly produce invisibly small particles – "pangenes", or "gemmules" – which travel to the site of production of sex cells, into which they are incorporated. A pangene is capable of giving rise to an organ exactly like that from which it has come, so alteration of the condition of that organ during the individual's lifetime can be passed on to its offspring thereafter. In some instances the pangenes might remain latent for a generation or two – which is why inherited characteristics can "skip" generations.

Assuming that Darwin believed that pangenes travelled through the bloodstream towards the sex-cell production centres (a reasonable assumption), Francis Galton (1822–1911) in 1870–71 carried out repeated blood transfusions on groups of differently coloured rabbits, hoping to show pangenesis at work. He failed. Faced with this, Darwin replied that he had never *said* the pangenes sailed the bloodstream, but omitted to explain just how they did make their journey. Despite his negative results, Galton continued to use the concept of pangenes while discarding the rest of Darwin's pangenetic theory – by now, belief in the inheritance of acquired characteristics was already fast waning – and came up with a theory remarkably close to current genetic ideas. Galton thought aggregations of pangenes could form entities – "stirps" – each stirp coding the individual for a particular characteristic. If for "pangene" one reads "nonoperational part of the gene" and for "stirp" one reads "genotype", one can see that Galton was not too far wide of the mark. (Mendel's results were not to be widely accepted until 1900.)

Another variation, intracellular pangenesis, was proposed in 1880 by the Dutch scientist Hugo de Vries (1848–1935) – who was later to be largely responsible for the publication of those researches by Mendel. Here the nucleus of each cell contains countless pangenes, which become active only when they pass from the nucleus out into the surrounding cytoplasm. This neatly explains why an individual's cells and organs change during his or her lifetime.

Darwin himself had earlier produced a theory of evolution by monads, a notion that came to him in 1837. This was an attempt to meet two theoretical requirements: that species adapt, and that old species die out to "make room" for new ones, so that the overall number of species stays much the same. He proposed that simple living forms, "monads", arose by Spontaneous Generation and instantly began to evolve as a new ancestral species. A monad – which can be viewed in one way as this ancestral creature, in another as a sort of species ladder – did not live forever. Eventually all of the species descended from the original would become extinct, leaving room for new ones developed from other monads.

Another evolutionary idea of Darwin's was that of "perpetual becoming"; it was a modification of the monad theory and dates from about the same time. Here a species (monad) will die out if it stops generating new species, will survive if it continues to produce viable offspring-species. This was a rather hamfisted attempt to explain extinctions in the fossil record while at the same time explaining the converse, survival when other species have failed. If you turn the whole thing over on its back, you arrive at something rather close to reality – as, indeed, did Darwin in due course.

PARALLEL EVOLUTION

The fossil record is littered with examples of parallel evolution, instances where creatures have evolved in such a way as to fill similar ecological niches, and thus have developed similar forms; likewise, the organs of species widely separated both genetically and in time often display striking similarities. For example, the marine reptile the ichthyosaur, which died out over 65 million years ago, was not unlike the modern dolphin, which is a mammal.

But just how far can parallel evolution go? The matter became one of importance in the middle of the 20th century, when various human fossil finds in the Americas showed anomalously ancient dates. Where before it had always been assumed that Man had been born in a single cradle-land, an Eden somewhere in Africa, the tantalizing possibility briefly emerged that perhaps *Homo sapiens* had appeared independently in several distinct cradle-lands at around the same time. Possible examples of parallel evolution in our species' fossil history, such as *Australopithecus*, tended to buttress the notion. Had it been true, it would have been excellent fodder for the racists. Moreover, the implications for geneticists would have been staggering: how could the separately evolved races breed, as it is perfectly evident they can? The answer proved simpler than any parallel-evolution hypothesis. Our species reached the Americas much earlier than anthropologists then realized.

It is worth noting that in the related field of cultural history it is accepted that parallel *cultural* evolution does take place. Despite the efforts of diffusionists, most famously Thor Heyerdahl (1914–2002), it seems likely that, for example, the Egyptians and the South Americans developed the idea of the monumental pyramid quite independently – the two styles of pyramid are anyway rather different. Should we ever encounter an intelligent extraterrestrial lifeform, we may well find that there are some similarities between their culture and ours, simply because, where they have confronted problems similar to our own, they may have opted for similar solutions.

EOLITHS

Most of us have seen prehistoric flint implements, and a surprisingly high proportion of us have actually found them. For a long time they were thought to be natural objects – thunderbolts, possibly – and people who suggested they might actually be artefacts dating from a time when men did not know how to fashion metals were likely to be laughed at. Still, it was rather difficult to explain them. Ulysses Aldrovandi (1522–1605) proposed that the objects owed their origin to "an admixture of a certain exhalation of thunder and lightning with metallic matter, chiefly in dark clouds, which is distilled from the circumfused moisture and coagulated into a mass (like flour with water) and subsequently indurated by heat, like a brick".

In 1707 John Frere (1740–1807), talking of flint objects which he had unearthed, at last took the plunge and insisted they were artefacts from a primaeval age (see page 46). In so doing, he vastly increased the recognized antiquity of *Homo sapiens* and of the Earth. Frere's postulation about the antiquity of the eoliths was at first vehemently rejected, but, once the principle that they were indeed ancient artefacts became current, the floodgates opened. In the mid-19th century an English grocer named Benjamin Harrison began to amass a vast collection of flints, which were christened "eoliths". These were of an intermediate stage between recognizable flint artefacts and common or garden lumps of flint: they were ancient artefacts if you sort of screwed up your eyes and crossed your fingers. The great advantage was that they were devastatingly easy to find – much easier than arrowheads – and so naturally enough eolith-hunting became a popular hobby. Professional archaeologists were bombarded with the myriad discoveries of the amateur rock-sleuths.

Surprisingly, some of the archaeologists don't seem to have minded, for in 1899 the Royal Society asked Harrison to display his eoliths for the benefit of its fellows. But there was always a nasty hard core of sceptics who maintained that any marks which the eoliths might show were probably the result of water action – the flints having been jostled together (with other stones, of course) in streambeds and the like. Harrison's collection was, at least from an archaeological viewpoint, quite valueless.

DEVOLUTION

A radical alternative to Darwinian evolution was heralded by Michael Cremo and Richard L. Thompson in their book *Forbidden Archaeology: The Hidden History of the Human Race* (1994) and brought to fruition in *Human Devolution* (2003), which is by Cremo alone. Both authors are members of the International Society for Krishna Consciousness, and therefore bring a "Vedic perspective" to their view of archaeology. Like the Christian Fundamentalists, they contend that archaeologists and palaeoanthropologists tend to suppress discoveries and information that conflicts with the accepted history of life on Earth, but they are considerably more scholarly in their approach and somewhat more rigorous in their evaluation of evidence.

As one example of the type of suppressed results they describe, there is the case of the discovery in the early 1970s by a team of US archaeologists at Hueyatlaco, near Puebla, Mexico, of some stone tools. When geologists from the United States Geological Survey dated the site, their various dating methods indicated it was some 250,000 years old. This was far older than the archaeologists could conceive, and therefore the datum has, Cremo and Thompson claim, simply been ignored ever since. Similarly, they cite the field research done in the 1940s by the Geological Survey of India in the Salt Range Mountains, now in Pakistan. The team discovered fossil evidence of plentiful flowering plants and insects in one stratum they investigated. The only problem was that the stratum appeared to date from the late Precambrian/early Cambrian, some 600 million years ago, long before, according to the accepted model of the Earth's past, either flowering plants or insects came into existence. The obvious explanation, that later strata had been buried under earlier ones by a geological thrust at some stage, was discounted by many of the researchers on the spot, who maintained there were no signs whatsoever of any such thrust having taken place.

So far so good, but then Cremo, in *Human Devolution*, invoked the Intelligent Design fallacy of "irreducible complexity" (he seems to have been seriously misled by Michael Behe's claim in *Darwin's Black Box* [1996] that no scientific papers had ever appeared modelling intermediate structures in the development of complex biomolecular structures) and the loopier paranormal researches conducted toward the end of his life by Alfred Russel Wallace (1823–1913), codiscoverer with Darwin of the principle of natural selection. Cremo is likewise injudicious in, elsewhere, pointing to the involvement of the Curies with the dubious Italian medium Eusapia Palladino (1854–1918) as an example of an important datum filtered out of the standard histories of physics. Cremo's purpose in drawing our attention to such items is to back up his assertion that the "human organism is composed of the elements matter, mind, and consciousness (or spirit)", that mind is capable of existing and operating independently of the physical organism, and "capable of acting on ordinary matter in ways we cannot explain by our current laws of physics". Needless to say, we are next treated to discussions of out-of-body experiences and past-life memories.

But then Cremo's thesis begins to get interesting again, even though it continues to rely too much on the kind of stuff you find in Pauwels and Bergier's *The Morning of the Magicians* (1960) as well as the point that the Universe appears – astonishingly! – to have been tailored to accommodate us. He posits the existence of a hierarchy of consciousnesses, with us on – of course – the bottom rung. Rather than having evolved upward from lesser creatures through the billennia of palaeontological time, we have *devolved*:

> Originally, we are pure units of consciousness existing in harmonious connection with the supreme conscious being. When we give up our connection with that supreme conscious being, we descend to regions of the cosmos dominated by the subtle and gross material energies,

> mind and matter. Forgetful of our original position, we attempt to dominate and enjoy the subtle and gross material energies. For this purpose we are provided with bodies made of the subtle and gross material energies. These bodies are vehicles for conscious selves. They are designed for existence within the realms of the subtle and gross material energies. Conscious selves who are less forgetful of their original natures receive bodies composed primarily of the subtle material energy. Those who are more forgetful receive bodies composed of both the subtle and gross material energies, with the gross material energies predominating.

Although DNA plays a part in this devolutionary picture, says Cremo, it is not an important part: our DNA merely codes for proteins, not for the way those proteins combine to produce a living organism. The key element in the production of an organism is, rather, a *bija*, or "mental seed", which we seem to derive not from our parents but from the reproduction of the demigods and demigoddesses who inhabit the plane of consciousness above ours. DNA and *bijas* thus act complementarily.

Within this model, there is no reason for human beings not to have been around on Earth from the earliest moment when the planet's environment became suitable for human habitation, which Cremo suggests was many hundreds of millions of years ago – indeed there's an imperative that this should have been the case, because otherwise there would have been little point in creating the terrestrial environment at all. (Along the way, humans have coexisted with the human-like creatures whom evolutionists regard as our hominid ancestors.) You might ask where the fossil evidence could be to justify such a claim. Well, first it must be recognized that, the further back one peers into palaeontological time, the less likely it is that one will come across any *particular* fossil; if humans were indeed on the planet several hundred million years ago, there's no reason we should expect to have turned up one of their fossils yet. Cremo suggests that in fact there have been evidences turned up of the longevity of the human strain, but that these have been swept under the carpet by archaeologists and palaeoanthropologists unwilling or unable to accept them

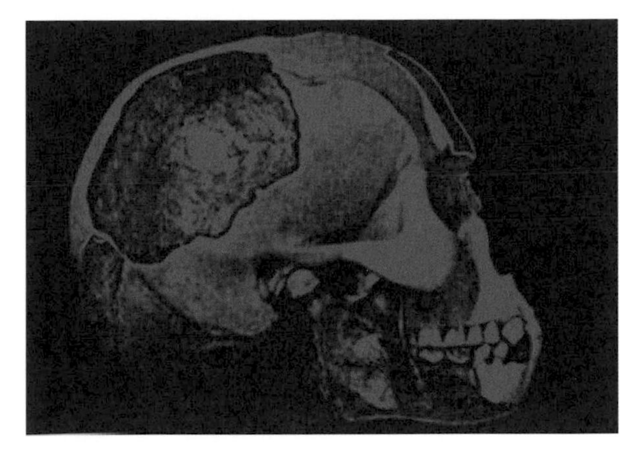

A hoax fossil hominid: Piltdown Man.

because they are so extremely in violation of the accepted history of *Homo sapiens*.

Another supporter of the concept that we've devolved from glorious ancestors is Jack Cuozzo, a New Jersey Creationist and author of the book *Buried Alive: Hidden, Suffocating from the Pain of a Story Left Untold* (1998). Cuozzo is a trained orthodontist, and became interested in Neanderthal teeth, examining them at various museums. He could tell no difference between Neanderthal and modern teeth except that Neanderthal teeth grew more slowly, and from this he concluded that the evolutionists must have things all wrong. The Neanderthals in fact emerged soon after the Flood, were the race to which the patriarchs described in the Bible belonged, and were lofty, civilized and intelligent folk who enjoyed remarkably long lifespans – again as recounted in the Bible. We are but degenerate versions of the Neanderthal ideal, one proof of our degeneracy being that we reach sexual maturity far earlier than they did.

Cuozzo is critical also of the belief by archaeologists and palaeontologists that their fossil finds in fact represent ancient humans. An example he cites is the Broken Hill Skull, discovered in 1921 in Kabwe, Zambia, by the Swiss miner Tom

Zwiglaar and described in *Nature* by Sir Arthur Smith Woodward (1864–1944), to whom it had been sent; it is regarded as the type fossil for a strain of *Homo erectus* called Rhodesia Man.

This first Rhodesia Man skull has a neat round hole just behind the left ear that could have been caused by a predator's tooth but is more likely a relic of trepanning (the piercing of the skull in an attempt to cure mental illness by releasing evil spirits. This explanation does not satisfy Cuozzo, who knows a bullet hole when he sees one. This is no prehistoric fossil at all but a modern human who was shot through the head. Of course, the skull doesn't *look* like a modern human skull – it's entirely the wrong shape – but Cuozzo can explain that too: the unfortunate individual suffered from acromegaly, a pituitary malfunction that can indeed produce a misshapen skull. Unfortunately for his argument, a skull's shape is one of the least important of the several means used to date it, and the others – which of course Cuozzo disputes – show that the Broken Hill Skull is at least 125,000 years old.

ANYTHING BUT AFRICAN GENESIS

Perhaps for racist reasons, it took a very long time indeed for theorists to acknowledge the possibility than humankind might have originated in Africa. From the time of the Enlightenment, when it began to be realized that the *Genesis* account of the Garden of Eden might not be literally true, people began looking around for a likely place where the first humans might have been created, and Asia – India especially – was a favourite spot. Of course, there was as yet little glimmering of the concept of evolution, so all that theorists really had to go on when expounded their hypotheses were cultural factors, notably language – those and their preconceptions. High mountains and a temperate climate were regarded as offering a suitable environment for the earliest humans; the realization that Indian and other Asian cultures were almost inconceivably old by Western standards was a further lure.

As the Enlightenment ebbed, the notion of humankind's

Asian birthplace continued to hold sway, reaching a new height in the German Romantic dreams of a noble, highly civilized, White Aryan race that originated in India and came to populate Europe, notably Germany. This myth-of-origin for the Germanic peoples was espoused, on linguistic grounds, by Friedrich von Schlegel (1772–1829) in such books as *Language and Wisdom of the Indians* (1808); and it was von Schlegel who coined the term "Aryan" in its meaning of "proto-Indo-European". A particular advocate, somewhat later, of the use of linguistics as a means of determining human prehistory was the racist French writer Joseph-Ernest Renan (1823–1892), author of *On the Origin of Language* (1858). Such notions were picked up by people like the French Comte de Gobineau (1816–1882) – author of such works as *Inequality of the Races* (1854) and *Moral and Intellectual Diversity of the Races* (1856) – and the UK-born writer Houston Stewart Chamberlain (1855–1927), not to mention the German composer Richard Wagner (1813–1883), and used by them to advance a racist, anti-Semitic, White-supremacist creed. With an admixture of the Nordics for no real reason other than that these thinkers admired the Nordic culture and mythology, the stage was set for the concept of the Aryan master race. (Much later, of course, the Nazis picked up the idea with even more horrific consequences.)

In the UK the idea of an Asian origin of humankind was far less popular, although some major figures, like Sir Charles Lyell (1797–1875) and Robert Chambers (1802–1871), subscribed to the hypothesis, albeit for scientific rather than patriotic reasons. The latter was eventually deduced to be the anonymous author of the two-volume *Vestiges of the Natural History of Creation* (1843–6), which advanced a rudimentary, somewhat Lamarckian hypothesis of evolution, which Chambers called transmutation. His concept of evolution concerned not just living things but *everything*: stars, planets, the Earth as well as living creatures had evolved from earlier models. Its anti-Creationism caused a scandal in UK society, of course, but more importantly was an influence on the thinking of Charles Darwin.

Of course, all of these Europeans who speculated about an Asian origin for humankind did so without finding any necessity actually to go to Asia and seek evidence for their claims. The one exception was the Dutch amateur palaeontologist Eugène Dubois (1858–1940). Unable to obtain any funding because of his amateur status, he "worked his passage" to Java by joining the Dutch East India Army, and after his arrival was permitted by the army to conduct his researches. There, in 1890, he discovered *Pithecanthropus erectus* – Java Man – which he believed was an intermediate form between ape and human: the Missing Link. When he presented the fossils and his hypothesis at an international zoological conference in 1895, however, the reaction was mixed, with most recognizing the significance of the fossils but declining to accept Dubois's hypothesis as to their nature. In the end Dubois, upset and frustrated by the controversy, became a recluse, taking his fossils with him. It was not until the 1920s that the US zoologist Ales Hrdlicka (1869–1943) coaxed the fossils out of him so that *Pithecanthropus* could properly be accepted into human ancestry. Dubois had in fact been wrong: Java Man was a specimen of *Homo erectus*, and thus fully part of the hominid lineage. Even so, the intimidation of Dubois is a shameful example of how theorists in a vacuum were prepared to dismiss the efforts of fieldworkers.

Stranger still were the circumstances surrounding Peking Man, also later identified as *Homo erectus*. In 1921 the first specimens were unearthed in some caves at a place called Dragon Bone Hill (the Chinese call fossils "dragon bones"), and they were sent for analysis to the distinguished anthropologist Davidson Black (1884–1934) at the Peking Union Medical College; he published his findings in *Nature*. The Rockefeller Foundation chipped in for further excavation, and from 1929 for the next seven years Chinese archaeologists, helped by Westerners such as Pierre Teilhard de Chardin (1881–1955), unearthed the remains of over 40 individuals. But then in 1937 the Japanese occupied the region, and excavations stopped. The fossils were stored for a couple of years at the medical college before the college's secretary, Hu Chengzi, despatched them to the US for safekeeping. Somewhere between the college and the port of Qinghuangdao they vanished, and they've never been recovered. Later researchers have therefore had to rely on the notes taken and the casts made at the time.

That the idea finally sunk in of humankind having its origins in Africa, not Asia, was initially thanks to the Australian anatomist Raymond Dart (1893–1988). While Professor of Anatomy at Witwatersrand University in South Africa, Dart came across fossil hominid skulls that seemed enormously older than any so far discovered elsewhere. He called the fossil species *Australopithecus africanus*, and published a paper in *Nature*. There was much skepticism in anthropological circles, primarily because "everyone knew" the first hominids had come from Asia; a secondary reason was that it was the received wisdom that the first adaptation distinguishing the human lineage from the rest of the primates was an enlargement of the brain, and the *Australopithecus* skulls indicated a relatively small brain. The Scottish anatomist Robert Broom (1866–1951) was convinced, though, and did extensive searches in South Africa for further *Australopithecus* fossils. Another to reckon Dart was right was the English anatomist Wilfred le Gros Clark (1895–1971), renowned as a man not easily taken in by extravagant claims: he had been among those to reveal the fraudulence of the Piltdown Man hoax. By the late 1940s the consensus was that *Australopithecus* was indeed a human precursor.

There are plenty of other conundrums in the story of the human lineage, even though the general pattern is now fairly well understood. For example, for decades it was believed that one of our earliest ancestors was *Ramapithecus*, a species known from various fragmentary fossils, the first of which was discovered in 1932 in the Siwalka Hills of northern India. *Ramapithecus* was dated to about 25 million years ago, which was thought to be around the time of the split between ape ancestors and hominids in the evolutionary tree. Comparative genetic studies between chimps and humans done from 1967 onward, however, demonstrated that this split must have happened as recently as about five million years ago, and in 1976 a complete *Ramapithecus* jaw was discovered, showing the creature to be definitely non-hominid. *Ramapithecus* is now known to have been extant about 12 million years ago – not the 25 million originally thought – and is generally classified in the genus *Sivipithecus* as a possible ancestor of the orangutan.

THE ACQUISITION OF VIRTUES BY CANNIBALISM

The idea that Neanderthal Man engaged in habitual cannibalism was born in 1899 when the Croatian anthropologist Dragutin Gorjanovi-Kramberger (1856–1936) found the remains of some 80 Neanderthals in a cave near Krapina; most of the bones betrayed signs of the bodies having been eaten, yet there were no traces of any predator having done the eating. The obvious conclusion, Gorjanovi-Kramberger decided, was that the eaters had been other Neanderthals, and cannibalism was an accepted part of the Neanderthal image for decades thereafter. By the 1980s, however, it was time for a re-examination of the evidence, and the consensus shifted radically: it was now assumed that the Neanderthals were not cannibals at all. However, further research done on the Krapina remains in the past few years suggests that, at least there, cannibalism was practised. It is feasible that cannibalism was a strictly localized habit among the Neanderthals, or it may have been just opportunistic; further evidence is awaited.

Less dubious is the conclusion that the Anasazi culture of what is now Colorado practised cannibalism around the 11th and 12th centuries. There are plentiful human bones that show signs of butchering and of being cooked, while analysis of feces found at the sites shows clear proof that the defecators had eaten human flesh. Whether this was included in the general diet or was part of some sacrificial ritual is unknown.

Eating human flesh and drinking human blood in order to absorb for oneself the qualities of the departed are practices which, while seemingly rare, may, then, be almost as old as *Homo sapiens*. Moreover, while the Judaeo-Christian tradition censures the practice, we should note that none of the other major religions have thought to mention cannibalism as a fundamental sin: if a person is dead already, what difference does it make? And even in Christian countries, in the Middle Ages, drinking the spurting warm blood from the neck of a just-beheaded felon was regularly practised as a cure for epilepsy; again in the Middle Ages, aphrodisiacs containing powdered human bones or flesh were popular.

The drinking of blood to increase fecundity and as an aphrodisiac goes back a long way. The tale is told that Annia Galeria Faustina, the wife of Marcus Aurelius (121–180), was so desperate to bear a son that she drank the warm blood of a gladiator who had "come second". This worked excellently. The resulting son, Commodus (161–192), was one of the most bloodthirsty rulers of the ancient world. Who said acquired characteristics can't be inherited?

The primitive equation of flesh and blood with strength was probably a result of observation. Dying people may waste away, losing flesh and strength together; or they may bleed to death, the loss of strength matching the loss of blood. Eating one's slain enemies probably arose from similar considerations. While it was a token of respect to eat the virtuous and strong people of your own tribe, eating a foe was a sign of ultimate hatred and revenge: not only had you killed him, now you were laughing at him as you stole his strength for your own use. In 1971 a member of the Palestinian terrorist organization Black September boasted proudly of drinking the blood of the assassinated Jordanian Prime Minister Wasfi Tel (1919–1971). In the later 1970s Ugandan dictator Idi Amin (1928–2003) was accused of having eaten the occasional organ of one of the dead in his torture camps. Whether or not he *did* is irrelevant: what is important is that his accusers had the idea in mind when they made the accusations. In some parts of New Guinea, the father of a newborn child will kill a family friend, and the victim's brains will be made into a ceremonial meal. The child then takes the victim's name. When the Western authorities tried to end this habit there was an outcry: how else could children get their names?

In *The Beginning Was The End* (trans. 1973), Oscar Kiss

Maerth claimed that humans first gained their intelligence (and enhanced their sexual potency) by eating the brains of their victims. Bands of prehistoric men roamed the land committing massacres and eating brains by the bucketful, and then had to carry out mass rapes because of the aphrodisiac qualities of their diet. Maerth painted a seedy picture of rings of dining clubs which he claimed were still in existence: now, however, the brains on the menu came from the higher apes.*

How much of all this is true? In *The Man-Eating Myth: Anthropology and Anthropophagy* (1979), William Arens surveyed a wide range of primary sources concerning cannibalism in the Caribbean, South America, New Guinea and West Africa, and found not a single eye-witness account: all were based on hearsay. He suggested that most accounts are merely propaganda, or of incidents of cannibalism as a last resort, as in remote aircraft crashes. However, there are extant eye-witness accounts from 13th-century Egypt and from China, where cannibalism was known at least until the 11th century, when some specialist restaurants were opened.

THE CLAN OF THE CAVE BEAR

The famous novel *The Clan of the Cave Bear* (1980) by Jean M. Auel (b1936), the first of her "Earth's Children" series, took the inspiration for at least its title from what was once accepted archaeological fact. During 1917–23 Emil Bächler, Director of the Natural History Museum in St Gallen, Switzerland, investigated a cave called Drachenloch in the nearby Churfirsten Mountains. He found various human artefacts of the Neanderthal Mousterian culture and also a large number of

* Maerth's book contains some other interesting hypotheses, as for example about sunstroke: a man exposed to the heat finds that his body uses so much energy trying to keep itself cool that it can no longer fight off the bacteria present in the bloodstream. The explanation as to why fish will never become intelligent is similarly curious: in order to become fully sapient, fish would have to be able to speak to each other; in order to speak they'd have to open their mouths; and, if they opened their mouths, they'd drown.

cave-bear bones laid out as if in ritual fashion; most prominent was a bear skull perched on top of a pile of other bones and with a leg bone put through its cheeks, as if the whole assemblage were some sort of shrine – Bächler described it as a "bone altar". This was clear evidence, he concluded, of religious activity on the part of the Neanderthals: at least locally there must have been a cult of the cave bear. Other palaeontologists searched for evidence of Neanderthal religious activities . . . and searched in vain. Further, when attention was focused on Bächler's excavation work it was found his methods were haphazard, to say the least; some of the results he reported were based on second-hand accounts obtained from the unskilled workers he employed. As for the skull with a leg bone placed through it, that could easily have been a matter of happenstance as scavengers fought over or otherwise threw around the bones of a dead bear. The notion of a Neanderthal cave-bear cult was a romantic one and took a long while dying, but is now generally disregarded as a fantasy – like Auel's novels.

AQUATIC MAN

The idea that our hominid ancestors went through a largely aquatic phase was broached by the distinguished UK marine biologist Sir Alister Hardy (1896–1985) in an article in the magazine *New Scientist* in 1960; he was not the first to do so, a precursor having been the German scientist Max Westenhöfer (1871–1957), who suggested much the same in his book *The Unique Road to Man* (1942). The notion has never in fact been discarded, but has been somewhat sidelined as a sort of wildcard hypothesis that's almost impossible to prove one way or the other. The UK writer Elaine Morgan has kept it in the public eye with a series of books, including *The Aquatic Ape* (1982) and *The Aquatic Ape Hypothesis* (1997), but of course popular books do not, and should not, influence the mainstream of scientific thought.

Hardy noted that the human body has some features that are rare elsewhere among the mammals, and certainly among our fellow-primates. For example, we have remarkably little bodily hair, and what we do have doesn't lie in a head-to-toe direction but tends to spread outward from the midline, as for example on the male chest. We can hold our breath, an ability almost unique in the animal kingdom. We have a single layer of subcutaneous fat all over the body. We walk upright. And so on.

Very few mammals are almost hairless, like we are, and almost all of them spend large parts of their lives in water or have had ancestors that probably did; hairlessness enables faster swimming – which is why competitive swimmers so frequently opt for depilation. Even without going that far, the pattern of our hair makes for fractionally more efficient swimming. The fat layer – well, people often jokingly call it blubber, and that term really indicates the thinking in the context of an aquatic human ancestor. The ability to consciously hold our breath is clearly invaluable when in the water, especially when hunting below the surface; it's hard to think why the ability would have been an advantageous adaptation for a creature who lived almost entirely on the land.

And then there's the human upright stance. The image of our ancestors emerging from the forests onto the plains and thereafter developing bipedalism, and in due course undergoing an explosion in brain-power, is increasingly being questioned by palaeontologists. What seems a more likely scenario is this: Perhaps seven million years ago the world saw the beginning of the ice age that ended (if indeed it has ended) just a few thousand years ago. Even tropical regions were affected: part of the global climatic change was a dying-off of great areas of jungle and the appearance in their place of grasslands. This necessitated a change in diet of many jungle dwellers, who found their traditional diet of fruits severely depleted, so that instead they had to do their best with grasses and bushes. The ancestors of the modern elephant and rhino were among those to take to the grasslands, as can be detected from their fossil teeth, which show adaptation to the tougher foodstuffs. Hominid ancestors, by contrast, it seems at first remained in the diminishing jungles, becoming more efficient fruit-gatherers: they developed bipedalism so that they could walk among and along branches with their hands free to grab fruit. By the time their exodus from the jungle became inevitable, they were already bipedal, or nearly so – a trait that gave them a strong survival advantage.

That is at least the conventional version. But no one has yet put forward a completely convincing explanation as to why our ancestors adopted this mode of locomotion.* Hardy, and Morgan after him, pointed out that there's one lifestyle in which an upright stance is not only easier for a habitually all-fours-moving creature initially to sustain but also could be of considerable survival advantage. That lifestyle is where a creature spends much of its time in relatively shallow water. The water buoys up the body, making it easier to stand on two legs only, while the uprightness means the creature can go further out from the shore or riverbank without the exertion and water-disturbance of swimming – for example, in search of food – while still keeping its head above the surface. They suggested that our ancestors went through a phase in which this was exactly their lifestyle. Later, when environmental changes encouraged our ancestors to transfer their activities primarily back to dry land, the habitual upright stance would likely have continued, the body having adapted to it by then; thereafter, the body would further have adapted for most efficient running and walking. This would explain the peculiarly "non-human" gait of the fossil hominid Lucy, who seems to have been a biped who walked in a way that was neither truly

* More may be learned by further study of a Kurdish family found in southern Turkey in 2005 who, apparently through a bizarre combination of their parents' genes, are naturally quadrupedal. Of the five brothers and sisters, only two are able to walk upright, and then only for short periods and with difficulty. Their quadrupedal gait differs from that of chimps and bears in that, rather than walk on the knuckles of their hands, they walk on the palms, with the fingers stretched upwards – as our remote ancestors may have done in order to leave the fingers free. Some – not all – of the investigating scientists think the behaviour may be an atavism, and that study of the unfortunate siblings may reveal vital information as to how our ancestors moved from a four-footed to an upright posture.

human nor truly ape-like: by her time our ancestors hadn't yet adapted fully to bipedal locomotion on land.

This is all very persuasive. The trouble is that it is as yet an unfalsifiable hypothesis: all we have of our pre-human ancestors are fossilized bones and, by the nature of the fossil record, we are given only intermittent cameos of pre-human development. The fact that we can find no unambiguous evidence of our ancestors having gone through an aquatic phase is indicative neither one way nor the other about the hypothesis – despite the dismissive conclusions of a 1987 conference on the subject held by the Dutch Association of Physical Anthropology, published as *The Aquatic Ape: Fact or Fiction?* (1991). On the other hand, that same lack of evidence makes the hypothesis an unnecessary one – a status that would of course rapidly change should unequivocal evidence one day appear. There's at least a one-million-year-gap between the time of our divergence from the other primates and the first known fossil hominids, a timespan that leaves plenty of room for us to have gone through an aquatic phase.

From the 1990s, the aquatic ape hypothesis was modified to produce what is often called the aquatic-hybrid ape hypothesis (AHAH): rather than being fully aquatic, our ancestors lived on the shores of lakes and seas and spent much but not all of their time wading in the water. This would explain, the AHAH proponents argue, the curious stance of Lucy.

Recently the AHAH version aquatic-hybrid ape hypothesis received a tremendous fillip from the work of Stephen Cunnane, Canada Research Chair in Brain Metabolism and Aging, and author of *Survival of the Fattest* (2005). He argues that anthropologists tend to underestimate the importance of the amount of nutrient the human brain requires for its functioning. In particular, the brain of a newborn baby consumes no less than 75% of the infant's energy supplies. Much of this comes from baby fat, present in plentiful supply in the newborn human (about 14% of bodyweight, on average). Other primates do not display this feature, so there's an obvious question as to how humans acquired the characteristic. Cunnane's answer is that our ancestors adopted a shoreline lifestyle in the

Great Rift Valley, with its plentiful lakes, rivers and wetlands. Such an environment offered a year-round diet of fish and shellfish, not to mention nutritious amphibians like frogs. That some australopithecines did indeed enjoy such a diet has been evidenced in studies done by Kathy Stewart of the Canadian Museum of Nature: our ancestors apparently found catfish especially succulent. With this vastly increased, readily and regularly available food supply, the way was open for baby fat to start appearing. Baby fat, as well as offering energy, is especially rich in DHA (docosahexaenoic acid), a polyunsaturated fatty acid crucially important to neuron function.

Further, this diet supplies plentiful iodine, to both babies and adults. Concentrations of iodine are far higher in aquatic than in terrestrial food sources. Iodine-deficiency is a major problem in the modern world precisely because we're adapted to a much higher iodine intake than is offered by our current diet, which for many people contains little or no aquatic food at all and for perhaps most contains not enough. (It's for this reason that in many countries there's a legal requirement to add iodine to table salt.)

It's a rule of thumb in evolution that adaptation arises in response to environmental pressure. In Cunnane's scenario, of course, the adaptation – humans developing bigger and better brains – would have been in response to an environmental *improvement*. However, rules of thumb are made to be broken: simply because adaptation is usually a product of adversity does not necessarily mean it has to be. The matter could instead be viewed as one of our ancestors adapting to fill a new ecological niche that had become available to them.

CREATIONISM

It should be stated at the outset of any discussion of Creationism that one thing hard for critics of – and indeed possibly supporters of – the 21st-century religiously motivated Creationists is to appreciate how far divorced these are from their 19th-century precursors, even those who were responding directly to Darwin in the decade or two immediately following

On the Origin of Species (1859). While the modern Christian Creationist is likely to insist on an age for the Earth of 6000 or so years, perhaps as much as 10,000 years, that the Earth and all thereon (humankind included) were created in six 24-hour days, and that the fossil record can be explained as evidence of all the species annihilated in the Flood, which was a global rather than a localized phenomenon, most of the Christian anti-evolutionists of Darwin's own time were prepared to be much more flexible in their view.

Matters were complicated by the fact that, just as there were many Christian spokesmen who were prepared to accept large chunks of evolution wholesale, reaching a happy compromise between the new theory and the *Genesis* account, there were also a few scientists who, for non-theological reasons, inclined toward the Creationist school; they accepted the principle that evolution had shaped modern life, but were essentially incredulous that natural forces alone could account for three things: the origin of matter, the origin of life, and the origin of humankind. At least three creations, they reasoned, must therefore be required, whoever or whatever might be responsible for the acts of creation.

Foremost among these scientific critics of Darwin was Louis Agassiz (1807–1873), Professor of Geology at Harvard, one of the great figures of modern palaeontology and whose Glacial Theory established that the past held a great ice age. His other hypotheses as to what had happened in the past, both to our planet and to its lifeforms, however, were less distinguished. His Creationist hypothesis was more complex than most, postulating not one or three but many creations. He envisaged a very traumatic history of the Earth, with life being wiped out many times over, each time being re-created to start the whole evolution process again. He accepted the fossil record as being what it seemed to be, but did not accept that the extinct organisms revealed therein were necessarily related to any of the organisms living today.

Those who attempted to bridge the gap between science and theology – to produce something that was purportedly a completely scientific hypothesis yet did not violate a liberal

Our ancestor as glorious savage. Engraving by Emile Bayard, c1860.

interpretation of *Genesis* – were scientists like the US geologist Arnold Guyot (1807–1884) and the Canadian geologist John William Dawson (1820–1899). Guyot very largely accepted the concept of evolution by natural selection, but insisted on the intervention of the divine for the three special creations: of matter, life and humankind. He put forward the notion that the six days referred to in *Genesis* were not really days in any normal sense of the word but instead six long eras. This concept became highly popular among both Christians seeking compromise and scientists unable to let go of the idea of divine intervention. Dawson accepted all of this with the exception of the concept of there having been only three acts of creation: he believed there must have been more than that. Neither man reckoned the Flood to have been more than a localized event.

For the first couple of decades after the appearance of Darwin's theory, most of the religious opponents to it – at least, the Protestants – were content to rely upon the scientific "disproof" of it by Agassiz and others. This means that they accepted the antiquity of the Earth, the non-universality of the Flood, and indeed a fair amount of evolution; they were content to regard the opening chapter of *Genesis* as something to be *interpreted*, rather than a literal account. This attitude persisted with most of them even when it became evident the scientific argument was being lost, when their arguments against evolution began to focus on its evident inconsistencies with theistic doctrine. But for some Creationists such accommodations were impossible, most notably the Premillennarians, of whom William Miller (1781–1849) had been the foremost precursor.

The Premillennarians were convinced that the end of the world was at hand, complete with Second Coming. Miller himself had a somewhat embarrassing career in this context, confidently predicting the Second Coming for 1833 and then, when that failed to materialize, for 1844, which proved another damp squib. He died before the publication of Darwin's *On the Origin of Species*; one can't help but feel it was a mercy for him that he did. He and his successors based their Creationist convictions on their own readings of the scriptures, and so had a very good reason to be antithetical to any theory that demanded the Earth be of considerable age, that the Creation had lasted more than six days, that the Flood had not been global, and so on: if any one part of the Bible could be shown to be other than literal and accurate, then so might other parts of it be – and this would imply the demolition of all their devout calculations.

Even so, many Christian Creationists were willing to concede that the *Genesis* account allowed for untold aeons of time to have passed between the creation of the Earth itself, along with most of its lifeforms, and the events of the Garden of Eden. This "Gap" hypothesis – referring to the gap between the fifth and sixth days, or during the sixth day – was, then, the second of the two compromises with the findings of geology that the Christian anti-evolutionists were prepared to make, the first of course being the belief that a "day" could mean a period of millions of years.

There were, however, exceptions to the rule of compromise. Curiously, however, these scriptural fundamentalists tended to belong not so much to the 19th century as to the newly dawning 20th. Perhaps it's something to do with the aftermath of the turning of centuries, times that are especially popular for predictions of the end of the world. Afterwards there must be bitter anticlimax among those who'd confidently anticipated the destruction of all, and a signal that they must redouble their efforts to counter the appalling sins of rationalism. The same seems to be true of the cusp between the 20th and 21st centuries. Herman C. Hanko, reviewing *Green Eye of the Storm* by the Creationist John Rendle Short for the *Standard Bearer* in February 2001, attempts to explain the extraordinary worldwide influence Darwin's "false" theory has had on human culture. It is not

> because of the scientific excellence of his theory. It has had to be revised more than once. The reason, I suggest, is because the theory destroyed the trustworthiness of the Scriptures, and especially the very foundation of the gospel in the first chapters of *Genesis*. And above all because Darwinism abolished the need for God and the

> Christian verities. Thus certainty was swept away. Nothing on the earth or in the sky could be guaranteed any more; everything was in a melting pot. Reality was nowhere to be found.

This stubborn clinging to the irrational in the teeth of the unwelcome facts presented by reality is known as *belief perseverance*, and is widely observed within the pseudosciences: where the facts contradict the belief, the facts are dismissed rather than the belief modified or discarded. A similar effect was noticeable among the Bush-loyalist section of the US population during the 2000s, where a high percentage continued to believe Saddam Hussein had possessed weapons of mass destruction and had been allied to Al-Qaida terrorists years after there had been copious public proof that neither was the case.

Why is it that the idea of evolution by natural selection conjures up a fury in adherents of the fundamentalist Christian Right that other scientific theories don't? The Big Bang Theory comes in for a fair amount of impassioned attack, but equally central theories like Relativity are almost if not entirely ignored. In December 2005 one of the most prominent scientific critics of ID, the US philosopher of science Daniel Dennett (b1942), was asked exactly this question by Jörg Blech and Johann Grolle for the German newspaper *Der Spiegel*, and replied:

> I think it is because evolution goes right to the heart of the most troubling discovery in science of the last few hundred years. It counters one of the oldest ideas we have, maybe even older even than our species. . . . It's the idea that it takes a big fancy smart thing to make a lesser thing. I call that the trickle-down theory of creation. You'll never see a spear making a spear maker. You'll never see a horseshoe making a blacksmith. You'll never see a pot making a potter. It is always the other way around, and this is so obvious that it just seems to stand to reason. . . . So the idea of a creator that is more wonder-

> ful than the things he creates is, I think, a very deeply intuitive idea. It is exactly this idea that promoters of Intelligent Design speak to when they ask, "Did you ever see a building that didn't have a maker, did you ever see a painting that didn't have a painter?" That perfectly captures this deeply intuitive idea that you never get design for free. . . . [Darwin] shows, hell no, not only can you get design from un-designed things, you can even get the evolution of designers from that un-design.

There's another possible explanation, though. Surveys in the US have repeatedly shown a strong correlation between level of education – particularly science education – and the disbelief in superstitions of all kinds, religion included. The single category of scientist with the highest level of atheism comprises the biologists, at 41% according to a recent poll. No wonder the Creationists see fit to attack education in general, science education in particular, and, most particularly of all, the bedrock theory of the biological sciences as opposed to that of, say, cosmology.

Meanwhile, the US Creationists' vilification of Darwin continues unabated, almost as if he were the Antichrist, such is its vitriol. Saturation abuse is not an uncommon weapon used by the US *faux*-Christian Right in their attempts to silence those against whom they know the rational argument to have been lost, but it does seem bizarre that they should sustain this campaign against a man who's been dead for well over a century. Here is Creationist broadcaster Ian Taylor ("The Demise of Charles Darwin", www.creationmoments.org):

> Reporting on Darwin's funeral, the *Guardian*, a tabloid of the High Anglican Church, said, ". . . lest the sacred pavement of the Abbey should cover a secret enemy of the Faith . . ." Christianity may rejoice in Darwin's burial at Westminster as a visible sign of, ". . . reconciliation between Faith and Science." In retrospect this statement is seen to be self-denial of prophetic words, because without doubt the sacred pavement does hide a secret enemy of the Church. That enemy has not only destroyed the faith of thousands from his grave but according to Scripture, renders spiritually unclean even those who pay respects to his grave.

CREATION SCIENCE

In the early 1970s US Creationists coined the new phrase "Creation Science" – essentially they were still peddling the same old Bible-based Creationism but had removed most of the overt religious references and added in their place a smattering of pseudoscience. Several factors contributed to the change.

One was the publication of the influential book *The Genesis Flood* (1961), by John C. Whitcomb Jr (b*c*1925) and Henry Morris (1918–2006), who began their discourse with the statement (in the second printing) that "the basic argument of this volume is based upon the presupposition that the Scriptures are true". It was primarily Morris's job to cobble together some kind of scientific or pseudoscientific rationale to support this assertion. Re-introducing the favourite Creationist refrain that scientific conclusions are merely a matter of interpretation – similar to the "it's only a theory" argument the ignorant or disingenuous use against evolution – he then presented a rehash of the revised geological scheme first offered by George McReady Price (1870–1963) in *The New Geology* (1923). The fossil-bearing rocks had almost all been deposited during the Flood, with the progression of the fossilized organisms from primitive to more advanced forms being a matter of differential buoyancy: basically, some creatures could swim better than others. Besides, the geologists were all wrong about the ordering of the stratigraphic column: was there not a site in Glacier National Park where Precambrian rock rested atop Cretaceous rock? The geologists attempted to explain such phenomena by invoking the concept of thrusting, but this was just special pleading: no one had ever seen thrusting at work. The notion of there having been multiple ice ages was just plain silly: there hadn't been enough time for more than one. The early times of the Earth had seen a clement environment because a big water canopy in the skies had created a greenhouse effect while also blocking off the harmful rays of the Sun; it was the collapse of this canopy that caused the Flood. And so on, and on, and on. Consequent upon the report by Clifford L. Burdick (1894–1992) that fossil human and dinosaur footprints could be found together in the bed of the Paluxy River in Texas, Morris claimed that humans and dinosaurs had co-existed.

A digression on the subject of dinosaur/human footprints: If the Earth's history has been extremely short, and if Man was created on the same day as the other animals, then it is inescapable that humans and dinosaurs must have lived alongside each other, just like in *One Million Years BC* (1966). In reality, of course, the dinosaurs became extinct about 60 million years before the first proto-hominids arrived, but this has stopped neither moviemakers nor Creationists from presenting the case otherwise. The greatest arrow in the Creationists' evidential quiver is the presence in the bed of the Paluxy River in Texas of well preserved human fossil footprints alongside equally well preserved dinosaur trails, the two sometimes overlapping in a way that showed they were almost certainly contemporaneous. And these aren't just ordinary human footprints: they're *giant* ones!

The unwitting culprit is probably one Roland Bird, who wrote about the footprints in a 1939 article for the magazine *Natural History*. Bird was careful to say that these were faked human footprints, manufactured by the locals for sale as souvenirs to tourists who'd come to see the perfectly genuine dinosaur ones. A pettifogging detail. The Creationists seized upon this "evidence" and, once it became clear the evidence wouldn't stand scrutiny, began to investigate the region in the hope of finding other, non-fake human footprints. They so far haven't been able to find any that are unambiguously human, but they have found some tracks that are, well, difficult to pin down: they could be human footprints that have been considerably eroded. Geologists have identified these as in some cases natural formations in the rock and, in others, severely weathered dinosaur prints, and reluctantly most Creationists have had to abandon the claim, although reserving judgement on the possibility that there might have been genuine human

Giant footprints from the Paluxy River bed.

footprints back in the 1930s when the subject was first mentioned but that these have since been weathered or eroded into invisibility.

At least until the past few years. New human tracks have been found! Scientists who've examined them say they aren't in fact human footprints, but scientists aren't to be trusted. One example came in 1968 when Creationist fossil hunter William (Bill) Meister, while investigating strata in Utah known to be 500 million years old, encountered what looked to him like a human shoeprint. Naturally this was seized upon by the less reputable Creationists as a proof that conventional geologists had it all wrong about the geological timescale; in fact, the "shoeprint" was quickly shown to be the product of the fairly common phenomenon called spalling, whereby pieces of rock flake away from each other along specific fracture lines.

Returning to *The Genesis Flood*, the Paluxy River data were quietly removed from the third revised printing, after it had become evident that Burdick had been overexcited in making the original report.

None of the arguments in *The Genesis Flood* are of course

science: this is a book of pure pseudoscience, comprising a mixture of wild guesses and straightforward fantasy. But to the uneducated reader it could look sufficiently scientific to disguise the fact that all Whitcomb and Morris were really doing was serving up the same old supper of God-created-the-world-in-six-days-about-6000-years-ago. The authors – and their countless supporters – were able to present this as a turning of the tables on science: where the trend had been to reinterpret the Scriptures in the light of each new wave of scientific discovery, now science was being reinterpreted in order to conform to the Scriptures. Whitcomb and Morris could portray themselves as being twin Davids combating the Goliath of the monolithic scientific edifice . . . and everyone loves an underdog.

A precise definition of the term "Creation Science" is hard to find. In his definitive survey of the topic's history in the US, *The Creationists* (1992), Ronald L. Numbers cites a 1981 Arkansas statute concerning the "balanced" teaching of Creationism and Darwinism which probably comes as close as any:

- Sudden creation of the universe, energy, and life from nothing.
- The insufficiency of mutation and natural selection in bringing about development of all living kinds from a single organism.
- Changes only within fixed limits of originally created kinds of plants and animals.
- Separate ancestry for man and apes.
- Explanation of the Earth's geology by Catastrophism, including the occurrence of a worldwide Flood.
- A relatively recent inception of the Earth and living kinds.

A further milestone in the story of Creation Science came in the early 1980s when Wendell Bird (b1954), director of the Institute for Creation Research, put forward, in conjunction with Paul Ellwanger, President of Citizens for Fairness in Education (CFE), the Theory of Abrupt Appearances. This hypothesis claimed that new organisms did not develop from older ones but appeared on the scene abruptly: one moment there's no such thing as a frog, and the next they're all over the place. It's hard to imagine how this "theory" could be regarded as scientific: the improbability of any such event ever happening would be the first strike against it, except of course that such branches of science as quantum physics are forever dealing in improbable events; but far more significantly the scenario it presents is in direct contradiction to all the evidence offered to us by the fossil record, where transitional forms abound. (A common tactic of Creationists is to say that no such transitional forms are known.* They will point out that no "missing link" exists between organism A and organism B. When organism C is discovered and neatly fills the gap, the Creationists will complain that there are no transitional forms between A and C or between C and B. And so on.) Bird's and Ellwanger's claim to science for their hypothesis seems to rest solely on the fact that they were careful to keep out of it any mention of divine intervention: the abrupt appearances just happened, that was all.

What the pair also brought to their campaign was the great new concept of "fairness": if the Darwinian theory of evolution was to be taught in the schools, then fairness dictated that so also should be their Theory of Abrupt Appearances. The "fair-

* Although it's a frequent Creationist mantra that, if evolution were real, surely there should be an abundance of transitional forms visible in the fossil record, it was a point that bothered the early evolutionists as well, T.H. Huxley included: Darwin's Bulldog was uncomfortably aware that he was Bulldogging in favour of something for which the evidence was slight. Huxley's doubts were laid to rest, however, during a visit in the 1870s to the US palaeontologist Othniel Charles Marsh (1831–1899), when he viewed Marsh's collection of horse fossils. Marsh had gathered and placed in order a more or less complete sequence of these, and it was evident that this showed a steady progression from ancient forms to modern ones. Here were transitions aplenty. That was well over a century ago.

ness" argument is of course completely specious, in that it implies the two schemes have equal weight. In point of fact, one is the keystone of all modern biology, tying together countless strands that were hitherto poorly understood at best, and has withstood innumerable tests for a century and a half, and the other is an untested, unevidenced hypothesis of the sort that people toss out in bars and later wish they hadn't. This concept of "fairness" would soon see *every* halfwitted and/or religious explanation of life, the Universe and everything being taught in schools. Someone thinks the Moon is made of green cheese? Right! Throw it into the syllabus. Nonetheless, it has re-emerged frequently in the debate over the equally spurious hypothesis of Intelligent Design, most famously in a 2002 pronouncement on the subject by President George W. Bush (b1946).

Of course, attempts to make Fundamentalist Creationism seem scientific go back far earlier than the 1980s. The standard of some of the Creationists' "scientific" arguments against Darwinian evolution can perhaps be assessed from this short extract from *The Evolution Of Man Scientifically Disproved* (1928), by William A. Williams (from Part One: The Evolution of the Human Body Mathematically Disproved):

> The population of the world, based upon the Berlin census reports of 1922, was found to be 1,804,187,000. The human race must double itself 30.75 times to make this number. This result may be approximately ascertained by the following computation:–
>
> At the beginning of the first period of doubling there would just be two human beings; the second, 4; the third, 8; the fourth, 16; the tenth, 1024; the twentieth, 1,048,576, the thirtieth, 1,073,741,824; and the thirty-first, 2,147,483,648. In other words, if we raise two to the thirtieth power, we have 1,073,741,824; or to the thirty-first power, 2,147,483,648. Therefore, it is evident even to the school boy, that, to have the present population of the globe, the net population must be doubled more than thirty times and less than thirty-one times. By logarithms, we find it to be 30.75 times. After all allowances are made for natural deaths, wars, catastrophes, and losses of all kinds, if the human race would double its numbers 30.75 times, we would have the present population of the globe.

Now, according to the chronology of [William] Hales, based on the Septuagint text, 5177 years have elapsed since the flood, and 5177 years since the ancestors of mankind numbered only two, Noah and his wife. By dividing 5177 by 30.75, we find it requires an average of 168.3 years for the human race to double its numbers, in order to make the present population. This is a reasonable average length of time.

Moreover, it is singularly confirmed by the number of Jews, or descendants of Jacob. According to Hales, 3850 years have passed since the marriage of Jacob. By the same method of calculation as above, the Jews, who, according to the Jewish yearbook for 1922, number 15,393,815, must have doubled their numbers 23.8758 times, or once every 161.251 years. The whole human race, therefore, on an average has doubled its numbers every 168.3 years; and the Jews, every 161.251 years. What a marvelous agreement! We would not expect the figures to be exactly the same nor be greatly surprised if one period were twice the other. But their correspondence singularly corroborates the age of the human race and of the Jewish people, as gleaned from the word of God by the most proficient chronologists. If the human race is 2,000,000 years old, the period of doubling would be 65,040 years, or 402 times that of the Jews, which, of course, is unthinkable.

INTELLIGENT DESIGN

The latest incarnation of Creationism, Intelligent Design (ID), can reasonably take as its starting point the book *Darwin on Trial* (1991), written by an ex-lawyer and born-again Christian, Phillip Johnson (b1940). His target was not so much evolution *per se* but the growing secularization of US society, for which he of course blamed school education; the teaching of evolution was merely the most prominent target. In the book he proved to his own satisfaction – and that of hundreds of thousands of scientifically untrained readers – that the theory of evolution by natural selection was not itself scientific but was a religion like any other save for its lack of acceptance of the supernatural. This is not to say that he did not believe evolution actually happened: he was quite prepared to accept evolution, but not the underlying principle of natural selection – i.e., the operation of random processes. Evolution, he said, was, rather, conducted under the guidance of God. Although as we've seen this is a centuries-old idea, it was enough of a springboard to bring into being the ID movement. As with Creation Science, ID bases its public platform almost to exclusion on attacking supposed flaws in the Darwinian theory rather than on advancing its own scientific alternatives. Its one (pseudo)scientific postulate is the notion of "irreducible complexity".

The basic notion of ID in fact goes back a long way before Johnson. The great physical chemist Robert Boyle (1627–1691), best known for Boyle's Law regarding the behaviour of gases, presented a vaguely similar notion in his book *Origin of Forms and Qualities* (1666): the Universe was a machine brought into being by God. A more significant precursor of the modern ID speculators was the English naturalist St George Jackson Mivart (1827–1900), who, in books like *On the Genesis of Species* (1871) and *Man and Apes* (1873) accepted evolution but not the mechanism of natural selection. A devout Catholic, Mivart was able to reconcile the evidence for evolution with his religious beliefs by saying that the adaptations of evolution must be designed and steered not by any random chance but by God. His opinion was based on a fallacy that is often offered by modern ID supporters as strong evidence of their claims: an adaptation did not become a survival factor until it was completely formed, so it must be created thus – and who could do this but God? The arguments Mivart put forward in favour of this scheme were sufficiently sound that Charles Darwin took the trouble in later editions of *On the Origin of Species* (1859) to include a section specifically rebutting them.

Mivart was far from alone in also finding the soul a sticking point. No less than Alfred Russel Wallace was another: Wallace said the human mind could not be explained by natural selection alone, and therefore must be the invention of God. Among the important US evolutionists, George Frederick Wright (1838–1921) said much the same. Their doubts were based on the fact that at the time no one could tell what the human mind actually *was*. Clearly it was seated in the brain, but equally clearly it could not be simply explained away as a brain function. This lack of comprehension was not simply because the "invention" of neuroscience was still a long way in the future; it is really only in the past few years that we have begun to come close to some understanding of the physical basis of the human mind, and the evolutionary processes responsible for producing it.

The fundamental proposition of ID, the notion of "irreducible complexity", is that certain biological components do not make sense as randomly evolved structures: until complete and in appropriate place within the appropriate larger structure they are useless; there must therefore have been a purpose in mind *before they started developing*. This purpose the ID proponents ascribe to a Creator, whom they are careful not to identify as God in their quest to have ID taught in US school science classes, which would be forbidden under existing laws separating Church and State if the hypothesis were to be overtly religious in nature.

A popular analogy they choose to promote is that of the mousetrap. A basic mousetrap consists of a piece of wood, a hoop of stout wire, a strong spring, another piece of stout wire designed to hold the hoop in place until the trap is sprung, and so on. None of these components has any utility whatsoever until the trap is assembled: all are designed specifically with the finished trap in mind. Missing from the logic is the understanding that structures can develop with one function – or simply be randomly occurring pieces of flotsam that have never been enough of a liability to have been, through adaptation, lost – and be put to use for another function. In the average garage are to be found lengths of sturdy wire, springs, bits of wood, etc., none of which were ever intended to be parts of a mousetrap. Put all of these together with countless other pieces of detritus in some technological analogue of the primeval ocean, subject them to a few tectonic or other external forces from time to time, wait long enough, and eventually a mousetrap will emerge – along with countless other structures of greater or lesser utility. One of these latter may in fact be a better, or at least differently, designed mousetrap than any we have so far devised. In fact, if you particularly wanted a mousetrap of the conventional design, you would almost certainly have to wait very much longer than if you were to be satisfied with *any* mouse-killing gadget. The wait would be even shorter for the first *useful* gadget to emerge – whatever the use might be. In other words, it is only by looking at the conventional mousetrap *after* it has appeared that the chances of its having done so randomly seem impossibly small: you are *imposing* the long odds by specifying a precise design of mousetrap.

Such recognitions were of course omitted from the "Bible" of the ID movement, the book *Darwin's Black Box: The Biochemical Challenge to Evolution* (1996) by Michael Behe (b1952). A qualified biochemist – he was Professor of Biochemistry at Lehigh University, Pennsylvania – Behe argued from that standpoint that numberless biological structures possess an "irreducible complexity": like the pieces of the mousetrap, they have no functionality until they're all brought together in the correct configuration.

What is most sinister about the whole ID movement is that its intent goes far beyond mere science; it is in fact a political movement. The prime mover behind the ID campaign is the avowedly right-wing Discovery Institute, founded in 1996. However much Creationists and ID proponents may try to laugh this off as rationalist paranoia or conspiracy theory, the fact remains that there is in existence a strategy document known as the Wedge which was indisputably compiled by Creationists at the grandiosely named Center for the Renewal of Science and Culture in the late 1990s, under the aegis of the Discovery Institute, which spells this out in detail, even in its preamble:

> The proposition that human beings are created in the image of God is one of the bedrock principles on which Western civilization was built. Its influence can be detected in most, if not all, of the West's greatest achievements, including representative democracy, human rights, free enterprise, and progress in the arts and sciences.
>
> Yet a little over a century ago, this cardinal idea came under wholesale attack by intellectuals drawing on the discoveries of

> modern science. Debunking the traditional conceptions of both God and man, thinkers such as Charles Darwin, Karl Marx, and Sigmund Freud portrayed humans not as moral and spiritual beings, but as animals or machines who inhabited a universe ruled by purely impersonal forces and whose behavior and very thoughts were dictated by the unbending forces of biology, chemistry, and environment. This materialistic conception of reality eventually infected virtually every area of our culture, from politics and economics to literature and art.
>
> The cultural consequences of this triumph of materialism were devastating. Materialists denied the existence of objective moral standards, claiming that environment dictates our behavior and beliefs. Such moral relativism was uncritically adopted by much of the social sciences, and it still undergirds much of modern economics, political science, psychology and sociology.
>
> Materialists also undermined personal responsibility by asserting that human thoughts and behaviors are dictated by our biology and environment. The results can be seen in modern approaches to criminal justice, product liability, and welfare. In the materialist scheme of things, everyone is a victim and no one can be held accountable for his or her actions.
>
> Finally, materialism spawned a virulent strain of utopianism. Thinking they could engineer the perfect society through the application of scientific knowledge, materialist reformers advocated coercive government programs that falsely promised to create heaven on earth.
>
> Discovery Institute's Center for the Renewal of Science and Culture seeks nothing less than the overthrow of materialism and its cultural legacies. . . .

That scientists and rationalists should object to this plan for the demolition of truth is not surprising; that anyone who believes in democracy and freedom of thought should likewise be horrified on learning of it is likewise unsurprising; and the fact is also that many Christian theologians are horrified by it too, not just from a scientific but from a theological viewpoint. As of late 2005, ID had made an impact on only one of the US's many religious universities, the Southern Baptist Theological Seminary in Louisville, Kentucky, which had created a Center for Science and Theology to accommodate leading ID evangelizer William Dembski (b1960). This was after he had left

Baylor, a Baptist university in Texas, in part because the faculty members at Baylor were opposed to the teaching of ID there. Other Christian universities, where they include ID in the curriculum at all, do so only in a historical/sociological/philosophical context: they reject it as science, since their faculty members are unimpressed by ID's claims to be a science and anyway generally see no fundamental conflict between Darwinism and their faith.

Around the turn of the 21st century the Templeton Foundation, whose aim is to seek reconciliation between science and religion, did some financial sponsoring of conferences and courses to debate ID, then asked the proponents of ID to submit grant proposals for research projects the Foundation might sponsor. But, reported the Foundation's senior Vice President, Charles L. Harper Jr, "They never came in." If even the proponents of ID themselves can't come up with any possible areas of research that might help test their hypothesis, this gives the lie to their claims that it is scientific. Or is it simply that they're *uninterested* in experiment and research, accepting the principle of ID as an article of faith without need for such encumbrances as proof? Rather like, say, a religious belief . . .

One criticism of Darwinian evolution often made by ID proponents (and orthodox Creationists) is that, if evolution by natural selection is a reality, why don't we see any evidence of evolutionary changes going on all around us? In fact, there's plenty of such evidence (Darwin's original Galapagos observations showed the effects of evolution over a relatively short period, which was what spurred him to propose his theory in the first place), but every time it's offered to the Creationists they raise the bar: that piece of evidence is somehow *not enough*.

What seems far too rarely to be done is to turn the tables and ask the Creationists and ID proponents to produce the living evidence of *their* hypotheses. After all, if new organs or new species are brought into existence ready-made, they should be popping up all over the place.

Where are they?

The conclusion of the celebrated 2005 trial in Dover, Pennsylvania, hailed as "the new Scopes Trial", over whether or not ID should be stressed – perhaps even taught – in the Dover schools, represented a triumph for rationality. US District Judge John E. Jones III, issuing his 139-page judgement on December 20 2005, effectively demolished ID as a purported science. He stressed he was not saying that ID shouldn't be discussed at all in schools, but that, even though its proponents "have bona fide and deeply held beliefs which drive their scholarly endeavors", "our conclusion today is that it is unconstitutional to teach ID as an alternative to evolution in a public school science classroom". Furthermore, "The citizens of the Dover area were poorly served by the members of the Board who voted for the ID policy."

Judge Jones was blunt also about the dishonesty of the pretence whereby a religiously based scheme was being dressed up as something scientific: "We find that the secular purposes claimed by the Board amount to a pretext for the Board's real purpose, which was to promote religion in the public school classroom. . . . It is ironic that several of these individuals, who so staunchly and proudly touted their religious convictions in public, would time and again lie to cover their tracks and disguise the real purpose behind the ID Policy."

Truth in science is not something that can or should be decided in a court of law: reality does not obey court orders any more than it does democratic votes. At the same time, it cannot be denied that the Dover trial did much more than settle a particular legal case: the doctrine of ID was effectively shredded, and very publicly at that.

This has, of course, not stopped other School Boards across the US from advancing plans to introduce ID into their science curricula. And in early 2006 senators in Utah were trying to enact legislation to ensure that schools in that state taught children about the "several" theories of life's origins. As the January 24 editorial in *The Daily Herald*, Utah, deriding the proposed measure as just yet another attempt to force religion into the science classroom, tartly concluded:

> Mostly, however, we believe all this is a colossal waste of time. Our legislators should spend their limited days on Capitol Hill doing something that will make a real difference to Utah.

PANSPERMIA

In 1743 the French naturalist Benôit de Maillet (1656–1738) suggested that germs of life came to Earth from space; they fell into the oceans and in due course grew into fish and, later, amphibians, reptiles and mammals. But it was not until 1908 that the first rigorous formulation of the panspermic hypothesis appeared, from Svante Arrhenius (1859–1927) in his *Worlds in the Making* (1906). According to him, spores are present in vast quantities throughout the Universe, drifting in space, driven from star to star by the radiation pressure of light.

The proposal was not ridiculous. Spores were known to be able to withstand extremes of heat as well as conditions of vacuum, remaining dormant until their environment became more temperate. However, more recently it has been appreciated just quite how much hard X-radiation is floating around in space; and this the spores could not withstand unless extremely well protected. Moreover, the proposal didn't really solve the problem as to the origin of life. It was all very well to say that life on Earth originated when spores fell from space, but how did the spores themselves originate? Better and better models of the early stages of the evolution of life on Earth were put forward, and so the hypothesis progressively lost favour.

More recently, however, Fred Hoyle (1915–2001) and Chandra Wickramasinghe (b1939) have come up with an interesting variant of the panspermic hypothesis. It has been known for some while that gaseous nebulae contain organic molecules: one nebula, for example, contains the alcohol equivalent of enough neat whisky to fill the hollowed-out Earth 1000 times over (a unique justification for the space race). But spectroscopic analysis seems to reveal that there are also more complicated organic molecules, the polysaccharides, in these interstellar clouds. As cellulose is one of the polysaccharides, this information is quite striking.

Further investigation into the composition of interstellar clouds of gas and dust (nebulae) and the processes underway therein has likewise proved promising. Astronomers have been able to show that molecules called polycyclic aromatic hydrocarbons (PAHs) are common in the nebulae. These extremely stable molecules are widely found here on Earth, too – for example, in car exhausts. Most recently, they've been found in open space as well, not just in nebulae. Astronomers like Adolf Witt of the University of Toledo have been painstakingly building up catalogues of the PAHs that can be shown to be present in the nebulae and in open space, and have found really quite complex molecules, such as anthracene and pyrene. The hypothesis is that smaller, less stable organic molecules that form in the nebulae, sheltered from potentially destructive radiation by the environment there, build up into stabler, more complex molecules that can survive escape into open space.

Physicists such as Louis Allamandola have been recreating the contents and conditions of interstellar nebulae in vacuum chambers in order to predict other organic chemicals that might be found in the real thing. An exciting early development was that, under these conditions, when the "nebula" is irradiated with ultraviolet light (plentiful in space, of course), there are photochemical reactions within the ice particles there. These reactions can convert simple molecules – like those of water, ammonia, methanol and carbon monoxide – into more elaborate compounds that form tiny protective membranes, reminiscent of cell membranes and possibly acting as protectors of the more delicate molecules within. The researchers have also created amino acids in their artificial nebulae; amino acids are the building blocks of proteins.

Perhaps most excitingly of all, Allamandola and his team found that, if they replaced one of the carbon atoms in a PAH with a nitrogen atom – something that happens frequently enough in nature – the end-products of the stimulated chemical reactions looked a lot like various of the components of

DNA and RNA. Spectra obtained from space closely match the spectra of these nitrogen-containing variants.

Some meteorites contain what look like rudimentary fossil cells; and these meteorites are thought to have arisen as a result of the fragmentation of cometary nuclei. Since comets are, as it were, messengers from interstellar space, their composition might reflect those of the gaseous nebulae; moreover, the materials detected in the outer parts of comets are quite compatible with the cometary nuclei containing complex organic molecules such as polysaccharides. Add to this the fact that the interior of a cometary nucleus is probably a relatively cosy place, suitable for biochemical reactions, and it seems possible that the seeds of life, if not primitive lifeforms, might well evolve there. When the comets struck the primitive Earth, they would of course be releasing all this organic material, so starting terrestrial life on its way. (More recent passages through cometary tails have, claimed Hoyle and Wickramasinghe, initiated epidemics.)

In an e-mail interview with the website Space.com in 2000, Chandra Wickramasinghe laid out his scheme of how panspermia operates:

> Life got started on a cosmological scale including the combined resources of all the comets around all the stars in all the galaxies of the entire universe.
>
> Once started, the robustness of life . . . ensures its essential immortality. It survives and is repeatedly regenerated in the warm watery interiors of comets. The space between stars is littered with cometary debris, some of which contains the seeds of life.
>
> Comets arriving at the Earth from the 100-billion-strong Oort cometary cloud of our solar system brought the first life onto our planet some 3800 million years ago.
>
> Evolution of life on the Earth was directed by the continued arrival of cometary bacteria, bacteria which must still be arriving at the present time.

In 2004 a team of scientists from the Washington University, St

Louis, and from the Lawrence Livermore National Laboratory in California, led by Christine Floss, analysing dust particles collected by NASA from the stratosphere, found organic material that was apparently older than the Solar System. This material seems almost certainly to have formed in an interstellar nebula. At the time of writing, analysis is under way of the dust collected by NASA from the fringes of Comet Wild 2 in the expectation of finding similar primordial organic material there – and perhaps organics that are even more exotic. Small pieces of cometary material are constantly raining down through the atmosphere onto the Earth's surface.

Even if the idea of living microbes surviving the long journeys from one star to another is finally shown to be unfeasible, what about rock-bound journeys between the Earth and its (relatively) close planetary neighbour, Mars? Certainly there are some Martian rocks on Earth, blasted off the surface of the red planet by impacting meteors or by spectacular volcanic eruptions and eventually making their way here. Although the same process is more difficult in reverse – our planet's gravity is stronger so the impact would have to be bigger, and the rocky fragments would be travelling away from rather than towards the Sun – it seems certain that at least some terrestrial material has made the journey to Mars. If we find microbial life on Mars, the first task will be to make sure that it didn't originate here.

However, it's the Mars-to-Earth scenario that is currently attracting the most interest. It has been pointed out that, in the early days of the Solar System, Mars offered a more clement environment for the emergence of life than did the Earth: being a smaller planet, it cooled down earlier than the Earth did. We know, too, that Mars has gone through at least one stage in its geological history when it was warmer and wetter, and thus favourable as an abode for life. The possibility is that life never got independently started here on Earth at all, but instead originated on Mars and was transported here in the form of microbial pollution on space rocks. We may have been looking to the wrong planet for clues as to the origin and earliest development of life.

The idea of life arriving on Earth in microbial form via meteors or comets received a further lift thanks to the red rain of Kottayam. In the early hours of July 25 2001 there was a crack as of thunder in the skies over the district of Kottayam, near the southwestern tip of India. A few hours later, red rain began to fall. Over the next couple of months there were numerous other reports: sometimes there would be a shower of red rain, other times there would be a period of red rainfall in the midst of a shower of more orthodox rain; sometimes the rain would be just tinted with red, other times it was as red as blood. Physicists Godfrey Louis and Santhosh Kumar of the Mahatma Gandhi University, Kottayam, were interested enough by the reports to gather samples of the red rain and analyse them under the microscope to try to find out where the red colour was coming from.

What the microscope revealed was pretty astonishing. The red admixture to the rain comprised what seem to be biological cells, but lacking both nuclei and DNA. The two physicists have accordingly suggested that the initial "thunderclap" was in fact the explosion of a large meteor high above the ground; there is nothing controversial in this suggestion, and it's generally accepted as the most probable explanation. Less widely accepted is the hypothesis that the red organic bodies in the rainwater are microbes that were inside the meteor, and were scattered all over the local atmosphere, including into rain clouds, by the explosion. Again, the idea of material being scattered by the explosion is not controversial; it's the identification of the cell-like structures as alien to the Earth that's disputed.

At the time of writing, further tests are being done on the bodies to see if perhaps they *do* contain DNA but that Louis and Kumar missed it in their examination. If DNA is found, that would probably indicate a terrestrial origin for the structures, although even then an explanation is troublesome.

The principal non-alien-origin hypothesis offered seems just as improbable as the alien-genesis one: that the meteor exploded in the midst of a flock of birds or bats, more or less vaporizing the unfortunate animals. The structures do visibly resemble red blood cells; but red blood cells don't last very long in water unless the water's salinity just happens to match extremely closely the salinity within the cells. The red rains were credibly reported over a period of a couple of months, so red blood cells are an unlikely candidate.

But there is something more startling to come. Louis tried culturing the microbes in a variety of nutrient solutions, and discovered that they reproduce asexually by binary fission, as do many unicellular organisms . . . but these do so at temperatures of 300°C! This is so bizarre that, Louis says, he and Kumar didn't dare put it into their paper on the red rain in case it was rejected out of hand. If this result is reproduced, then the alien-origin hypothesis may be nearly invincible.

The implications of all these different lines of research are several. While no one can as yet speculate as to whether life itself can form in interstellar nebulae – perhaps a simple virus or bacterium – it seems fairly clear that life's building blocks can, and do. The arrival of these on the early Earth would have given a kick-start to the early development of life; this might explain why there was such a (relatively) short time between our planet becoming habitable and the emergence of the first self-replicating cells. But, if that happened here on Earth thanks to the omnipresent space-formed organic molecules, then inevitably it must have happened on all, or almost all, other potentially life-bearing planets as well. This implies that life may be much commoner in the Universe than we might currently expect – and that of course would be a triumphant vindication of Svante Arrhenius's original speculations.

INDIGO CHILDREN

According to some, the next stage of human evolution is among us . . . in the form of indigo children. You can tell indigo children from ordinary ones because their of high IQ, high intuitive abilities, high self-confidence . . . and high levels of attention-deficit disorder, except it isn't really attention-deficit disorder but yet another symptom that they're indigo children.

Oh, yes, and their auras are a distinct indigo colour, which is where they get the name.

Indigo children were first recognized as such in the 1970s by Nancy Ann Tappe, a San Diego parapsychologist, and were the subject of the book *The Indigo Children: The New Kids Have Arrived* (1999) by Lee Carroll and Jan Tober, which by early 2006 was said to have sold over half a million copies in the US, presumably to the suffering parents of children exhibiting what seemed very, very like attention-deficit disorder. And indigo auras.

The documentary movie *Indigo Evolution* was released in 2006.

According to Doreen Virtue, a former psychotherapist and author of books like *The Lightworker's Way: Awakening your Spiritual Power to Know and Heal* (1997), *The Care and Feeding of Indigo Children* (2001) and *Goddesses and Angels* (2005), "They're vigilant about cleaning the earth of social ills and corruption, and increasing integrity. Other generations tried, but then they became apathetic. This generation won't, unless we drug them into submission with Ritalin."

But wait! Even the indigo children aren't as evolutionarily advanced as the subjects of Virtue's book *The Crystal Children* (2003). Crystal children don't exhibit the same very-like-attention-deficit-disorder symptom as indigo children; indeed, ignorant parents might be concerned that their crystal children were, to put it politely, slow developers. However, these children are very fond of nature: "Doreen once watched one Crystal Child walking from tree to tree, giving each one a big hug."

Enigmatic rock carving from Tassili in the Sahara.

CHAPTER 4

Aliens Among Us

In 1609 Galileo Galilei (1564–1642) turned his primitive telescope towards the Moon, and immediately saw that it was a world. From that moment on, there was a delicious free-for-all between those who maintained that all other worlds must have civilized inhabitants and those who were frankly sceptical: the debate over the "plurality of inhabited worlds" was fiercest in religious circles. The fervour of the idea's supporters can be judged from the fact that both Dominique Arago (1786–1853) and William Herschel (1738–1822) maintained the Sun itself was the home of civilized life. As late as the 1950s the Russian technical magazine *Voprosy Filosofi* gave "proof" that every planet in the Universe is inhabited: dialectical materialism demanded this be so.

The debate has continued ever since. There are few today who would say that there are no other intelligent and even technologically advanced lifeforms in the Universe. This is a viewpoint distinct from those endorsed by many amateur theorists who still adhere to the view that, if it's a world, there must be somebody living on it.

Robert Goddard (1882–1945), the US "Father of Rocketry" – his Russian counterpart was Konstantin Tsiolkovsky (1857–1935) – was commendably sceptical about the then much-bruited theory that Mars was inhabited. He was, however, totally convinced that other stars have planets and that some of those planets bear intelligent life. Here, though, his imagination rather forsook him for, when he came to speculate as to how bizarre those star-beings might be, he could say only that "there may very likely exist human beings like ourselves, probably with strange costumes and still stranger manners".

Christiaan Huygens (1629–1695) believed that Jupiter's having four moons (only the four Galilean moons had then been discovered) betrayed a plentiful supply of hemp on the planet. His theory was born from the popular fashion of finding design in all parts of the Universe. Clearly, the only purpose of Earth's Moon was to act as a navigational aid. Since Jupiter had four moons rather than just one, this must mean there were a lot of seafarers there. Hence a lot of boats. Hence a lot of sails – and hence a lot of ropes with which to pull the sails up and down. And . . . to make a lot of ropes, you need a lot of hemp.

Distant Pluto has not escaped attention. Astronomers have long ago concluded that Pluto is probably one of the most sterile places in the Solar System, but some ufonauts tell us this is not the case. The planet is warmed by chemically produced light in its atmosphere (why its atmosphere is not frozen solid is unclear), and has inhabitants who are much like ourselves, only sexually perverted. Intriguingly, those same ufonauts omitted to mention Pluto's rather large moon Charon until astronomers found it in 1978 or its two further small moons until they were detected in 2005. This is part of a consistent pattern whereby aliens are ignorant of astronomical information until it's discovered by human scientists.

The science of SETI – the Search for Extraterrestrial Intelligence, generally using radio waves as a medium – is a relatively recent one, having come into being only in the latter half of the 20th century. So far it has detected no radio signals from civilizations native to another solar system, but then one would hardly expect it to have done so so soon. There are countless candidate stars to look at it in turn using large radio telescopes on which observation time is at a premium, and for each star the radio astronomers are relying on the extremely long-shot coincidence that we happen to be looking at Them just at the same time as They happen to be beaming signals in our direction. A better approach might be to look instead for side-effects of Their advanced technology: an interstellar propulsion drive, for example, could be detectable at a considerable distance. However, this latter approach is in its infancy as yet.

Some of the amateur theorists, of course, jumped the gun a little. We're told that two Russian astronomers, Valentina Zhurvleva and Genrikh Altov, believed radio messages from the star 61 Cygni were received on Earth in 1882, 1894 and 1908; they further suggested that the second of these messages was in response to the 1883 eruption of Krakatoa – the inhabitants of 61 Cygni naturally thought the bang was Earth's primitive way of responding to their first message. Since radio astronomy was not born until 1932, the claim seems doubtful; and it becomes even more so on learning that it comes from W.R. Drake's *Gods or Spacemen* (1964), especially since he evades the radio-astronomy issue by giving the impression that the two Russians were unsure as to whether the signals used radio or laser – the latter being, of course, even more anachronistic: how could anyone have been able to detect a laser signal in 1882, 1894 or 1908?

ANCIENT ASTRONAUTS

The basic thesis of the ancient-astronaut hypothesis is as follows:

In the remote past the Earth was visited by alien space-travellers; primitive human cultures regarded these visitors as gods and their advanced technology as miraculous. The stories of the "gods'" doings were passed down by word of mouth, being much corrupted in the telling; the dedicated researcher can unravel these distorted legends to give clear proof that the aliens really did walk among us.

The ancient-astronaut literature is enormous – and grew especially rapidly in the two decades or so following the English-language publication in 1969 of Erich von Däniken's *Chariots of the Gods?* To analyse the entire range of von Däniken's "evidence" would be impossible here, since even a cursory examination would fill an entire book. Indeed, several entire books have been devoted to exactly that, of which the best is possibly still *The Space-Gods Revealed* (1976) by Ronald Story.

Von Däniken was far from the first ancient-astronaut theorist. Among the most fecund before him was W.R. Drake, whose *Gods or Spacemen* (1964) must rank among the most resolutely implausible books ever published. In it Drake seeks to interpret semiliterally the mythical struggles of the gods Saturn, Jupiter, Uranus, etc. He proposes that, where the ancients refer to the god Jupiter (say), what they are really talking about are the inhabitants of the planet Jupiter. He is on less certain ground when he extends this argument to cover the Things from Uranus. Admitting that Uranus was not discovered (or "rediscovered", as he puts it) until 1781, he nevertheless maintains that it was a planet known to the ancients – at a time when, possibly, all the planets were much more closely bunched about the Sun. This speculation might sound reasonably defensible . . . until you realize that the *name* Uranus was not given to the planet until after its "rediscovery" by Sir William Herschel in 1781. Herschel himself wanted to call the planet Georgium Sidus.

The extraterrestrials slogged it out on Earth in prehistoric times using "terrible siderial [sic] and nuclear weapons". Drake points up the intriguing possibility that the Van Allen belts might be the radioactive detritus of this ancient nuclear war.

Another pre-von Däniken theorist was Desmond Leslie (b1921), co-author with George Adamski (1891–1965) of the bestselling *Flying Saucers Have Landed* (1953). He told us that astronauts arrived here from Venus in the year 18,617,841BC, a date which he ascertained after having deciphered "ancient Brahmin tables". Since the human race is at most about 4 million years old, it's hard to know where the Brahmins' date came from: who could have written it down?

Von Däniken himself seems to believe that, if one theory is reasonably convincing, then three alternative theories will together be very convincing. Take the matter of Man's parentage. In *Chariots of the Gods?* we are told we arose from breeding between hominids and visiting aliens; the genetic implausibility is obvious.* Perhaps because he, too, thought of this, von Däniken tells us in *Gods from Outer Space* (1970) that the aliens instead simply tampered with humanity's "genetic code" to make our ancestors more intelligent. Yet in *Gold of the Gods* (1973) there's yet another version. Extraterrestrials arrived on Earth in flight from a different group of aliens. "Our" aliens hid in a great complex of tunnels which can still be found – if you are Erich von Däniken – under Ecuador and Peru, having craftily placed on the Fifth Planet (see page 77) various bits of decoy equipment. The conquering hordes of attacker aliens, tricked by the ruse, blasted the Fifth Planet to smithereens and went on their way. Eventually "our" aliens crept out onto the Earth's surface – wearing breathing apparatus, as frequently depicted in ancient drawings – and became interested in the antics of a species of semi-intelligent ape, whose evolution they philanthropically spurred.

Von Däniken has personally visited those tunnels in which the extraterrestrials sheltered, in the company of one Juan Moricz; there he saw wonderful golden artefacts and the like. Alas, Moricz later claimed that he made no such expedition with von Däniken, who was accordingly forced to modify his account a little, stating that what had really happened was both (a) that he had indeed been in the caves with Moricz, but not in the main part, and (b) that he had *not* been in the caves, but

* Carl Sagan once observed that a visiting extraterrestrial would have less chance of impregnating an ancestral human than an ancestral human would have of impregnating a petunia: the petunia, coming from the same planet as the hominid, would be biologically closer.

had simply "fictionalized" his account a little to make it more exciting. This latter represents an avenue of approach which is too often neglected by orthodox scientists.

Von Däniken has much to tell us, too, about the aliens and the ancient Egyptians. For example, mummification is typical of the "superscience" we might expect visiting aliens to pass on to the natives: the plan was that the extraterrestrials should return a few millennia later to revive the mummified corpses. The process of mummification includes the extraction of the corpse's brain, a practice that must surely lead to all sorts of complications in any proposed revivification process.

The aliens also taught the Egyptians to build pyramids, helping out with liberal doses of "supertechnology". This help was vital, according to von Däniken, because the Egyptians had no rope (in fact they did have rope), no wood (they had access to abundant supplies of wood), not enough food to feed all those slaves (except that Egypt was the bread-basket of the ancient world), and no means whereby to cut the "bricks" from solid rock (the rock was limestone, and the tools the Egyptians used were simply made of harder stone – in fact, tool-marks are still actually visible in some of their quarries). How else, von Däniken asks with telling effect, could the ancient Egyptians have learnt to solve the architectural problems of pyramid-building? Perhaps the aliens helped out with the maths? For, as von Däniken has pointed out, if you multiply the height of the Pyramid of Cheops by a factor of one billion you get an answer of about 158 million kilometres, which is quite close to the mean distance of the Earth from the Sun, 150 million kilometres. Surely, he argues, this can be no coincidence.

Kenneth L. Feder, in his survey of archaeological pseudoscience, *Frauds, Myths, and Mysteries* (3rd edn 1999), gives a valuable insight into why so many of the illustrations von Däniken reproduces seem to show evidence of ancient technology and/or astronauts, even though such evidence is not plain until von Däniken has explained it to us. Feder describes the process as "inkblot" observation (the scientific term is "pareidolia"). The basis of the famous Rorschach inkblot test is to present an individual with a randomly formed inkblot and, through asking the person what s/he sees there, to find out what is going on in her/his mind. We all undergo the same sort of process when, for example, we look at cloud formations and "see" horses or castles. As Feder points out, we already *know* what's on von Däniken's mind: it's no wonder he sees ancient-astronaut evidence every time his eye falls on an image whose meaning is not immediately apparent.

Throughout von Däniken's work we are tantalized by curious little side-hypotheses. He makes much of the famous vision of Ezekiel (see page 221), the account of which is obviously a primitive attempt to describe a spaceship – or, according to former NASA engineer Josef Blumrich (1913–2002) in *The Spaceships of Ezekiel* (1974), a sort of helicopter, complete with four rotors. Returning to von Däniken, there is the suggestion that even Satan was an extraterrestrial visitor. Other writers have had similarly biblical thoughts, such as Gerhard R. Steinhäuser, with *Jesus Christ, Heir to the Astronauts* (trans 1974).

A further phenomenon which has absorbed the ancient-astronaut theorists concerns the mysterious Nazca lines in Peru. These are, we're told, the runways of an international or even interplanetary airport. Since all self-respecting UFOs have vtol facilities, and since if they didn't their wheels would get swiftly stuck in the soft soil of these "runways", it's hard to guess just what kind of ancient aircraft could have used this "airport". The "runways" are arranged not at all as one would expect them to be, laid out as they are in the form of animals, birds and geometric shapes. And where are the world's *other* alien airports? One airport on its own is not very useful.

Once such points had been made public, the ancient-astronaut theoreticians backtracked to what we can best regard as Plan B. The pictures drawn by the Nazca lines can be fully appreciated only from the air, so that the prehistoric Native Americans must have required alien help in order to both plot out the lines and appreciate the results of their labours.

Members of an organization called the International Explorers Society became interested in this problem, and wondered if perhaps the Nazcan Indians had invented the hot-air balloon. In 1975 two members of the society flew to a height of about 180m above the Nazcan plain in a balloon constructed from materials known to have been available to the ancient Nazcans. This was certainly high enough for the patterns of the lines to be appreciated; and, while the experiment of course did not prove the ancients had ever performed this exercise, it certainly undermined any concrete statement that they *could not* have done so. Besides, even if the Nazcans did not know how to fly over the plain, is there any reason to suppose they were unable to work out at ground level how to create such vast markings? Certainly the remains of models of the patterns have been discovered, so it seems plausible the ancients worked from these. But why should the Nazcans have gone to such lengths to create designs which they themselves could never look upon? One obvious speculation is that they did so for religious reasons: it mattered not at all to them that they could never properly appreciate their own work, because it was done for the gods – the sky is, after all, a traditional abode for deities. An alternative suggestion is that the lines comprised a colossal calendar/observatory, like Stonehenge, built on such a huge scale for purposes of permanence.

Yet another vital piece of evidence in the ancient-astronaut larder is the Piri Re'is map, dated 1513 and discovered in 1929 in Istanbul. At first sight it appears to show the coastline of Antarctica – that is, the coastline of the Antarctic continental landmass, today completely obscured by snow and ice. The map was discussed at great length by Charles Hapgood (1904–1982) in *Maps of the Ancient Sea Kings* (1966); he suggested there might have been a very early maritime culture whose boats occasionally sailed along the Antarctic coast at a time when it was free from ice. How their information was passed down to Piri Re'is, working in the 16th century, is not especially clear, but this need not necessarily invalidate Hapgood's speculation.

To many ancient-astronaut theorists, however, the map was first made tens of millions of years ago and shows the

Antarctic coastline *exactly* as it has recently been discovered to be. Direct comparison of the Piri Re'is map with a contemporary chart of the true Antarctic coastline unfortunately shows this latter claim to be quite simply false. It is possible, perhaps, that the extraterrestrials were simply bad cartographers, since the Piri Re'is map also shows such oddities as the tip of South America joined onto Antarctica and not one but two Amazon rivers. And the claim that they made their map by direct aerial survey 70 million years ago, when the dinosaurs were still roaming the Earth, seems to be invalidated by the fact that India is not shown in the middle of the Pacific Ocean, which is where the subcontinent was at the time.

All in all, it seems likely that Piri Re'is compiled his map from a number of sources, fudging the issue when he found inconsistencies in the data. Some of the information he received was quite good, some of it very bad. It is not at all inconceivable that, even as early as 1513, European explorers to South America had discovered there was a great landmass to the south, and that their cartographers simply put a wiggly line on the map to represent it: the history of cartography is littered with examples of mapmakers having acknowledged in this way lands about which they knew nothing beyond the fact that they existed somewhere in that general direction.

Before leaving this map, it is worth noting that in *The Morning of the Magicians* (1960) Louis Pauwels and Jacques Bergier tell us that Piri Re'is personally presented his map to the US Library of Congress.

A startling hypothesis is put forward by Craig and Eric Umland in their *Mystery of the Ancients* (1974). According to the Umlands, the ancient Maya were originally explorers from another Solar System. Their main base was on the Fifth Planet (see page 77), but when this exploded they had to make do with a settlement on Earth – although not before they had mined the core of the Moon. The reason the colonizing Maya made their base in the middle of the South American jungle was that they wished to avoid the possibility of catching diseases from nasty *Homo sapiens*: their bodily defences would not have been able to cope with our ancestors' home-grown viruses and bacteria. This puzzles. Even if they succeeded in isolating themselves entirely from contact with our ancestors, the Maya were still exposed to the countless other germs you find in jungles. Anyway, we discover from the Umlands, the Maya were, despite the precaution, in due course ravaged by venereal disease – a problem, incidentally, that does not seem to have afflicted von Däniken's miscegenating "gods".

We next find ourselves wondering why it is that such a supersophisticated species as the spacefaring Maya should indulge in such barbaric practices as regular human sacrifice. The Umlands supply us with the obvious answer: those were not human sacrifices at all but dissections, a distinction wholly misunderstood by the primitive human witnesses, possibly because it is confusing to be dissected while you are still alive. The Maya were – but of course! – performing all these dissections in order to try to do something about the venereal disease that was ravaging them. As for their habit of rudely plucking the hearts from the bosoms of the living, this was another woefully misunderstood practice. They were conducting heart-transplant operations.

A further problem that the Umlands have concerns time; in particular, they have a cavalier way with dates. Consider this: "Why the Maya chose the Antarctic section of Gondwanaland as their main base we may never know. But it proved to be a costly mistake. Between 40,000,000 years ago and the beginning of the Ice Ages, conditions must have continually worsened at the Pole." Well, perhaps not quite so foolish: the Maya would have enjoyed a mere *35 million years or more* of reasonable tranquillity before having to move.

Again in the geological context we find, concerning the Maya's habits of practising the ultimate in deep mining in search of heavy metals: "It is easy to see how the shifting continents would continually hamper the Maya in their attempts to tunnel through the Earth's outer crust and mantle to reach its

Eyewitness sketch done during the airship flap of 1896–7.

molten core." No sooner, one imagines, had the Maya managed to erect a drilling rig than along would come a shifting continent to bowl it over.

In his book *The Mayan Factor: Path Beyond Technology* (1987), José Argüelles adds to the Umlands' information on the Maya. The Maya were visitors from a planet of the star Arcturus. They came here not in spaceships but by transmitting their DNA in code form. Once here, they latched onto the Olmec, educating them in such arts as writing, which thereafter spread to the rest of humanity. At the moment they're on their way back for a second visit, with an ETA of December 21 2012;

on that date there will be some sort of tremendously spiritual transformation of reality, and our species' story will take an upward turn and never be the same again. Or something. In the meantime, it was extremely important for as many people as possible to go to holy sites on the weekend of August 16–17 1987; this "Harmonic Convergence" would indicate to the aliens that we were ready for their arrival. Of course, should the aliens fail to turn up on December 21 2012, Argüelles will always be able to argue that it's our own fault: not enough of us heeded his call to participate in the Harmonic Convergence.

Perhaps the most exciting moment in *Mystery of the Ancients* comes in the section titled "The Mayan Menace". The Maya who stayed at home on their own planet are coming, all these millions of years later, to rescue the ones stranded here; and, when they arrive, those newcomers will zap us. But what really sets the blood racing is the discussion stemming from some odd radio sources discovered by the Russians in the early 1970s, and at the time thought by them possibly to be signals from an extraterrestrial civilization. After that, the Umlands claim, both the US and Soviet governments engaged in a cut-throat race to try to decipher the ancient Mayan script.

Why?

Because those signals "would obviously be in Mayan".

Probably the only theory in the whole gamut of popular ancient-astronaut theories to be worth serious further consideration is that proposed by Robert K.G. Temple (b1945) in *The Sirius Mystery* (1976). Temple focused on anthropologists' reports of the beliefs of the Dogon, a North African tribe that had been comparatively little contaminated by Western culture, and made much of Dogon descriptions of elliptical orbits for the planets of the Solar System and for the Pup, the white-dwarf companion star of Sirius; the Pup was discovered telescopically only in 1862, although its existence had been deduced as early as 1834. The Dogon knew, too, that the Pup orbits Sirius in about 50 years. This seems fairly striking: it is a sophisticated enough piece of knowledge for a nontechnological culture that planets should go around stars in circular orbits; to know that those orbits are ellipses rather than circles requires not just a further veneer of sophistication but a quantum leap of the understanding. However, it has been counterproposed that all that this represents is the fact that the Dogon *like* the ellipse – they use the shape everywhere in their art – so that it would be as natural in their culture to assume elliptical orbits as it would be in ours to assume circular ones. This cuts both ways, of course: the reason for the importance of the ellipse in Dogon culture could be that they were taught basic astronomy by the Visitors, and thus realized the fundamental importance in Nature of the shape.

Among other pieces of Dogon knowledge/belief were that Jupiter has four moons (presumably the Galilean moons) and that Saturn has rings – both pieces of information that are impossible to obtain without the aid of a telescope. But there is a curious element of anachronism here: as we now know, Jupiter has many more than four moons, and the other giant planets, not just Saturn, have ring systems. Surely visiting extraterrestrials would have told the Dogon the truth of the matter, rather than what was state-of-the-art astronomy some while ago? (In fairness we should note that it is of course possible that the investigating anthropologists simply did not bother to record Dogon nonsense about Uranus and the rest having rings, regarding it as just a primitive fantasy, an extrapolation from the received knowledge about Saturn.)

That the Pup, the companion of Sirius, is a very small, compact, hot star was shown in 1915 by the US astronomer Walter Adams (1876–1956); the knowledge that Sirius A and B – as they are more formally called – orbit each other in a period of about 50 years dates back much further. The first recorded anthropological investigation of the Dogon – by Marcel Griaule (1898–1956) – took place in the 1930s and 1940s. Carl Sagan (1934–1996) and others have asked: could it not be that some explorer or missionary might have visited the Dogon before Griaule and, on being asked about Sirius – which to the Dogon is the most important star in the night sky, a view shared, for agricultural reasons, by the ancient Egyptians – told them of the latest scientific knowledge? Such information could then easily and very quickly have become a part of the rich tapestry of Dogon mythology.

But, if this is the case – and such things are known to have happened on a smaller scale and over a shorter timespan elsewhere – one wonders why the Dogon mythology does not contain elements of, say, Christianity. Why is there no mention of, for example, mythical wheeled monsters that make a great noise and travel faster than a man can run? This lack points up a serious flaw in the mundane explanation of what Temple was quite probably right to call a mystery. That said, at the end of the day the more prosaic scenario, involving a Visitor from the West rather than from the skies, does seem *simpler*.

Temple naturally mentions the legend of Ea, or Oannes, a strange being who, according to legend, brought civilization to the Sumerians during or just before the 4th millennium BC. Oannes had the overall form of a fish, but with an extra, humanoid, head beneath the piscine one, and feet as well as a tail. He came from the sea. According to a version by Alexander Polyhistor (*fl*70BC) of an account written down by the Babylonian priest Berosus (*fl*260BC), Oannes gave men

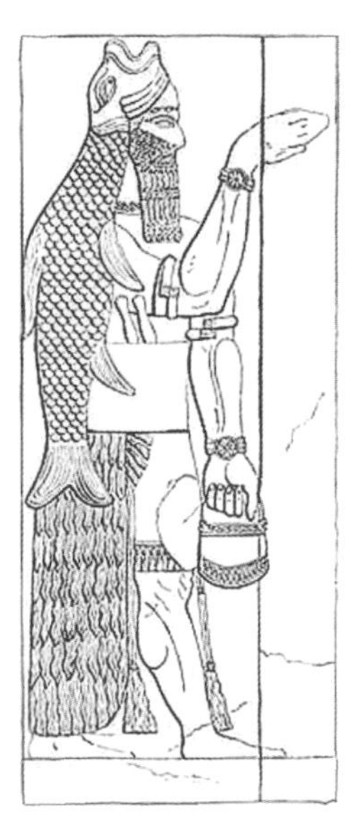

> an insight into letters and sciences and every kind of art. He taught them to construct houses, to found temples and to compile laws, and explained to them the principles of geometrical knowledge . . . In short, he instructed them in everything that could tend to soften manners and humanize mankind. From that time, so universal were his instructions, nothing material has been added by way of improvement.

From time to time thereafter, various of Oannes's fishy compatriots popped out of the sea to help the Sumerians on their way. Certainly Sumerian civilization did flourish with almost unnatural suddenness, coming from nowhere to achieve a high cultural and social level in a period of only a few generations, so the legend is far from contra-indicated by the archaeological evidence.

It is obvious to some that Oannes and his fellows were extraterrestrials operating out of a submarine base. Since they could not breathe our atmosphere, they were always seen clad in spacesuits whose helmets bore, no doubt, a fishy crest. This could have made them look as if they had two heads; alternatively, the one-head-above-the-other description might simply be a representation, according to artistic conventions other than our own, of one head *inside* another. Whatever the case, these extraterrestrials did not otherwise look very much like human beings. Although their bodies seem to have been of roughly the same length as ours, most of this length was dragged along the ground behind them. (It goes almost without saying that, despite such considerations, Oannes was, according to the Umlands, a typical Mayan.)

However, all of the three ancient accounts of Oannes that are still extant are based on a single original, the report written down by Berosus; but Berosus lived thousands of years after the event – much longer after it, indeed, than we are living after the time of Christ. Berosus may have been recording the popular mythology of his time, or he may equally have been a sort of ancient equivalent of Geoffrey of Monmouth (c1100–1154), whose fertile imagination provided us with the elaborations of the Arthur legend.

Even so, the descriptions of Oannes that we have are uncannily like accounts of mermen; and we cannot *totally* discount the idea that mermen and mermaids might be a corrupted folk memory of creatures like Oannes. One has to be careful of this sort of reasoning: it is just as likely that it was the mermaid myth which gave rise to the legend of Oannes.

What can be viewed as an ancient-astronaut variant of the Intelligent Design hypothesis (see page 182), with elements also of the panspermic hypothesis (see page 189), is the Intervention hypothesis promoted by the US psychology graduate and ex-US Military Intelligence officer Lloyd Pye, who promotes his theory with a charming mixture of slightly uncertain erudition and a wry, often self-deprecating humour. His notion, expressed in the book *Everything You Know is Wrong* (1998), is that all the marvellous creatures of the modern Earth could not have come into existence through random evolution: there must have been a designer somewhere along the line. However, unlike the ID proponents, he doesn't think this designer was supernatural and he believes the "intervention" was much more radical than a mere tinkering about with microscopic structures. He envisages an ancient starfaring alien species who, as our Solar System emerges, recognize it as a good place for life and accordingly "sprinkle" its planets with

> a variety of two separate forms of single-celled bacteria they know will thrive in any environment (the extremophiles). But the bacteria have a purpose: to produce oxygen as a component of their metabolism. Why? Because life almost certainly has the same basic components and functions everywhere in the universe. DNA will be its basis, and "higher" organisms will require oxygen to fuel their metabolism. Therefore, complex life can't be "inserted" anywhere until a certain level of oxygen exists in a planet's atmosphere.

About a billion years later the "terraformers" returned to drop a second batch of prokaryotes on the planets, and thereafter they made periodic visits to check on progress. Some two billion years after the start of this particular project they realized that life was flourishing only on the third planet from the Sun, so henceforth they concentrated their attentions on it; the only other planet where life had established a toehold was the fourth (i.e., Mars), but the prospects didn't look good there, so they abandoned it.

The young Earth by now had abundant land and liquid water, so it was time for the aliens to dose it with eukaryotes – larger and more complex than the prokaryotes, these are more fragile and require a more amenable environment, but they are capable of producing oxygen at a far higher rate. Some 1.4 billion years later, the aliens reckoned the Earth was ready for its first batch of multicellular organisms, the group of enigmatic (to us) organisms we know only as fossils and call the Ediacarans. And so the process continued, with the aliens dropping by every few tens or hundreds of millions of years to tend their Earthly garden.

> If, indeed, terraformers are behind the development of life on Earth, nothing else makes sense. If, on the other hand, everything that happened here did so by nothing but blind chance and coincidence, it was the most amazing string of luck imaginable. Everything happened exactly when it needed to happen, exactly where it needed to happen, exactly how it needed to happen. If that's not an outright miracle, I don't know what is.

Up to this point, Pye's hypothesis has not been unreasonable – improbable, perhaps, but not unreasonable. Here, however, he's invoking as supportive evidence some very dubious hidden assumptions. The improbability of these events having happened by chance is remarkable only if one considers the Earth in isolation. But the Earth is one of only untold billions of planets. After all, these things must have happened by chance *somewhere*, or the alien terraformers themselves could not exist. Second, there's no need to conjure aliens to bring along the eukaryotes and the Ediacarans when the Earth's environment was ready for them. Life evolves to fit new ecological niches when they become available, and the changing environment of the Earth was creating new ecological niches aplenty.

Pye claims the aliens are still making visits, although it's been a long time (in our terms) since the last one. During these visits they have performed various feats of genetic engineering on the lifeforms of Earth, tailoring them in accordance with an unknown plan. He makes much of the unusual fur of the cheetah, which is doglike over much of the body but catlike on the spots; clearly this is a product of cloning experiments. And then of course there's us: humankind is the focus of the most intense genetic-engineering attention of all.

There's also the matter of the "Starchild Skull". This was given to Pye by a Texas couple in 1998, and for some years now has been in a UK laboratory for analysis. If genuine and not a human artefact, it seems an obviously mammalian and probably primate skull, tentatively dated to about 900 years ago, and it does indeed look very strange; whether it belonged to an alien is of course another matter entirely. Pye reckons the scientific analysis is taking so infernally long because the scientists realize that it is alien and are terrified to admit it.

Pye's Intervention hypothesis is perhaps the most extensively worked out development of what used to be called the Bootstraps hypothesis of the origin of life. While we are slowly getting closer to an understanding of how life started, only a few decades ago it was not uncommon to hear evolutionary scientists dutifully mentioning this as a far-left-field but nonetheless quite serious rival proposal which could not be entirely discounted and which had the considerable virtue of avoiding the tricky organic and physical chemistry necessary in any attempt to plot the stages between a collection of inorganic chemicals, on the one hand, and a living organism on the other.

Imagine a "visitor" to the primaeval Earth. He lands, climbs out to have a look around, finds nothing of interest, returns to his spaceship, and goes. But: microorganisms from his boots have contaminated the ground on which he walked; or, alternatively, he emptied out the tank of his chemical toilet, leaving a puddle of bacteria-rich effluent. Either way, he has infected our world with life; and from these lowly microorganisms evolve, over the aeons, the lifeforms that we know, including ourselves. A version of it is wryly illustrated in Bruno Bozzetto's animated movie *Allegro non Troppo* (1976), done to the musical accompaniment of Ravel's *Bolero*.

THE TUNGUSKA EVENT

Exactly what caused the explosion that occurred in a remote part of Siberia just after 7am on June 30 1908 is not known. For various reasons the site was not properly inspected until nearly 20 years later, and almost certainly much evidence must have been lost. Investigators have had to rely on what was left of the site – spectacular enough evidence in itself – as well as, in the earlier days, on eye-witness accounts. The most probable culprit is a very large meteor, although the idea that it might have been the nucleus of a small comet is still current. In either event, the body involved exploded before hitting the ground.

According to those eye-witnesses, on that June morning an elongated object, brighter than the Sun and trailing clouds of multicoloured smoke, sped across the sky with an earsplitting roar to explode in the region of the Stony Tunguska River. There then appeared a pillar of fire, followed by clouds of black smoke that ascended to a height of some 20km. The explosion was indeed a mighty one, estimated to have been of the order of 30 million tonnes of TNT, which puts it on a scale with the largest hydrogen bombs. The shock waves in the atmosphere travelled twice around the world before dissipating, and seismic waves were detected as far away as Washington DC. For some time the aurora borealis was especially prominent, and magnetic disturbances in the atmosphere were detected several thousand kilometres away. Across much of Europe, high-altitude "silvery clouds" were observed, and luminosity in the night sky was for some time bright enough to cause panic as far distant as London; closer to the affected region, people could take photographs at midnight.

It was not until 1921 that a scientific expedition was sent out to investigate the region. This was led by Leonid Kulik (1883–1942), an expert on meteorites from the Mineralogical Museum at Petrograd; he had read the accounts and assumed the event had been the impact of a giant meteorite. The expedition was not able to visit the actual site, but everything Kulik discovered persuaded him he had been right in his assumption. He returned to the region with another expedition in 1927, and this time was able to reach the site. What he found was astonishing. Over an area of some 3000 square kilometres devastation had clearly reigned. Trees were scorched and, many showing signs that their branches had been flash-incinerated, thrown flat on the ground, their roots pointing inward towards what presumably had been Ground Zero. But at the centre of the site, as so clearly defined by the trees' direction of fall, Kulik could find no trace of the meteor crater he anticipated. Instead he found many relatively small holes a few metres deep and up to a few dozen metres across. Clearly, he reasoned, the meteor must have exploded in midair, the force of the blast driving fragments into the ground at colossal velocity. But, if that had indeed been the case, where were the meteorite remains that should be at the bottoms of the holes?

So, if it wasn't a meteor, what could it have been that had exploded in the Stony Tunguska region? The suggestion that it might have been a cometary nucleus emerged during the 1930s and has been the generally accepted scientific explanation ever since. Under certain circumstances a cometary nucleus could explode in midair and create the observed effects. Such impacts, common in the early days of the Earth, are fortunately very rare now that the Solar System has largely settled down. The last similar impact that we're fairly certain about is the one that wiped out the dinosaurs, some 65 million years ago.

However, with the first public demonstrations of the atomic bomb at Hiroshima and Nagasaki in 1945, many people started pointing out that the Tunguska explosion showed features of a nuclear detonation – for example, the mushrooming cloud of smoke immediately after the explosion. (It should be borne in mind that mushroom clouds are not unique to nuclear explosions. They are a characteristic of firestorms, which may be produced by non-nuclear means.) Much of the energy output of the explosion had been in the form of concussion, heat and light – again a characteristic of a nuclear blast. The temperature at the core was deduced to have been several million degrees; surely only a nuclear explosion could have produced such an intensity of heat? There was evidence pointing to the possibility of radiation sickness among domesticated animals in the region, and the local plant life, too, showed possible signs of genetic damage.

Soon after WWII the Soviet engineer Aleksander Kazantsev (1906–2002) suggested in a novel which he may have intended to be taken seriously that the explosion had been that of the nuclear-powered engine of a Martian spacecraft. At the time half the world was convinced of the reality of the Martian canals and the civilization that must have built them (see page 72). Then, in the later 1940s, it was proposed that the Earth had been struck by a lump of antimatter; however, such an explosion in the atmosphere would have resulted in the generation of a great deal of radioactive carbon; a survey made in the region found no evidence of this.

In 1973 Albert A. Jackson and Michael P. Ryan published a paper in *Nature*, "Was the Tungus Event Due to a Black Hole?", putting forward the notion that the Stony Tunguska region could have been struck by a high-velocity mini black hole perhaps the size of an atom. Such an event could indeed produce many of the effects observed after the explosion of 1908. For a long time, however, the hypothesis was flatly rejected on the grounds that there was no record of any corresponding convulsion shortly afterwards, elsewhere in the world, to mark the black hole's emergence from its journey through our planet. More recently, however, the Jackson–Ryan hypothesis has resurfaced, since it has been pointed out that there's absolutely no reason to believe a black hole penetrating the Earth need come out the other side: if its mass were to be appropriate, it could remain inside our globe, in a presumably precessing orbit around the Earth–Moon system's centre of gravity. If so, our planet is doomed: the black hole will continue to gobble up matter within the Earth, unknown to us until, sooner or later, its mass is sufficient for the effects to become only too horribly evident. An excellent novel exploring this concept (and much else besides) with some fair measure of scientific rigour is Bill DeSmedt's *Singularity* (2005).

In 1976 John Baxter and Thomas Atkins published *The Fire Gone By: The Riddle of the Great Siberian Explosion*, which was a speculative but scientifically fairly plausible exploration of the hypothesis that the event had been caused by a malfunctioning alien spacecraft. Baxter and Atkins pointed out that the eye-witness reports, when collated, indicate the object didn't take the simple southeast-to-northwest course generally assumed, but zigzagged. An interpretation of this erratic path could be that extraterrestrials aboard their doomed craft were desperately making course corrections so that when it exploded or crashed, as they knew it must inevitably do, it would do so in as remote and unpopulated a region as possible.

The doomed-spacecraft hypothesis has not died. In 2004 came a cluster of reports from Russia that fragments of such a spacecraft had finally been discovered in the Stony Tunguska region, by an expedition headed by Yuri Lavbin. The world waited with bated breath, but little more was heard. It was pointed out that fragments of spacecraft might indeed have been found there: the region is a drop zone for discarded stages of Russian rockets launched from the nearby Baikonur space base. Lavbin's own hypothesis was that a giant meteor was heading toward the ground but that beneficent extraterrestrials intercepted it and blew it up in order to prevent a global disaster.

What then about those putative course corrections? Meteors do not always follow a straight-line trajectory as they plummet down through the atmosphere: differential ablation of their leading surfaces can alter the path, or they can start breaking up, with different fragments following slightly different courses. Collation of the eye-witness accounts could easily have caused observations of several different objects moving with different paths to be interpreted as a single object making course changes. Remember, too, that everything was happening very swiftly indeed.

In the 1980s and 1990s, knowledge about the behaviour of large meteors and small cometary nuclei as they hit the Earth's atmosphere improved markedly, thanks to direct spy-satellite observation and progress in the art of computer simulation. A small cometary nucleus of the standard dirty-snowball model – which not all cometary nuclei may be – disintegrates very soon after hitting the atmosphere: the explosion happens tens of kilometres above the ground, and no one notices except Pentagon spy-satellites. Computer simulations indicate that the Tunguska body was almost certainly a stony meteor some 60m in diameter that exploded about 8km above the ground. But that may not be the last word on the subject.

UFOLOGY

A curious phenomenon worth noting at the outset is that very, very few UFOs are reported by amateur astronomers. In most of the developed countries, more man hours are spent annually staring at the night sky by amateur astronomers than by everyone else combined, so one might predict it would be from this relatively small section of the population that most UFO reports would emanate. In fact, exactly the opposite is the case.

There is no single "UFO hypothesis": ufology is a field of study which draws together hordes of original theorists and many disparate theories. UFOs may be (a) craft from outer space; (b) craft from other dimensions; (c) craft from the future; (d) natural phenomena, but of a type as yet unknown to science; or (e) various mundane phenomena misinterpreted by witnesses. Clearly (e) is the least romantic though most likely of these options, but (d) should not be forgotten. A Chicago dietician whose name is lost to us believed UFOs were merely hallucinations resulting from dietary deficiencies. He claimed that if everybody ate 50 dandelions a day the UFOs would disappear.

The great charm of the UFO hypotheses is that they link together so neatly with all the other popular theories. Perhaps UFOs come from Atlantis or Lemuria? According to the Aetherius Society, the Star of Bethlehem was a UFO. W.R. Drake, in *Gods or Spacemen* (1964), manages the best such link-

ur Killed In Cowlitz County Air Crashe

The Kelsonian-Tribune

ALLEN BLACKWELL, CI STUDENT PILOT, DAVE DAVIDSON VICTIMS OF FIRST AIRPORT TRAGEDY

U.S. ENGINEERS SET HEARING ON KELSO SPAN FOR AUGUST 29

Kelso Traffic Moving More Normally

Whoozit?

FLYING DISK INVESTIGATORS DIE IN ARMY BOMBER WRECK

Election May Be Asked In NK District

age of all. He tells us Tibetans maintain that parts of their country are periodically visited by alien spacecraft. These leave radiation trails in the snow, thereby giving rise to legends about the Yeti.

The scarcity of evidence in favour of the alien-spacecraft hypothesis might seem to put that hypothesis in doubt, but even here the ingenuity of the human mind has an answer. Ufologist John A. Keel (b1930) pointed out that this very scarcity is in fact a strong proof of the hypothesis: the aliens *must* be here, because who else would be covering up their tracks for them?

Surprisingly, Margaret Mead (1901–1978) was a UFO devotee, believing the lights in the sky were indeed alien spacecraft – although she diverged from most observers in thinking they could well be robots or drones "controlled from elsewhere in space". As to their purpose, she wrote (as cited by Henry Gordon in *Extrasensory Deception*, 1987): "The most likely explanation, it seems to me, is that they are simply watching what we are up to . . . keeping an eye on us to see that we don't set in motion a chain reaction that may have repercussions far outside our solar system."

In most modern works on the "UFO phenomenon" we read that the first sighting was by Kenneth Arnold (1915–1984) in 1947. Arnold in fact said the objects he saw were crescent-shaped, but that they moved like saucers (or flat stones) when you skip them across water. He thought they might be remote-controlled Soviet missiles, so on landing tried to get in touch with the FBI to pass on a warning. The local office of the FBI being shut, in desperation he went instead to the nearest newspaper office, that of the *East Oregonian*. It was *East Oregonian* journalist Nolan Skiff who, in haste – the deadline for the next day's issue of the paper was almost on top of them – amalgamated the two descriptions as "saucer-like". The description stuck. Not least, it influenced the aliens themselves, because, to judge by the other reports that started flooding in and have flooded ever since, they promptly redesigned their spacecraft to match Skiff's journalistic error.

More recently, with the sudden popularity of the ancient-astronaut hypothesis, it has become fashionable to mention also supposed UFO sightings in the Bible – especially *Ezekiel* 4 5-28:

> . . . a stormy wind blew from the north, a great cloud with light around it, a fire from which flashes of lightning darted, and in the centre a sheen like bronze at the heart of the fire. In the centre I saw what seemed four animals. They looked like this. They were of human form. Each had four faces, each had four wings. Their legs were straight; they had hooves like oxen, glittering like polished brass. Human hands showed under their wings; the faces of all four were turned to the four quarters . . .

In the rationalist *The UFO Enigma* (1977) Donald Menzel and Ernest Taves made a good case for this particular vision being of an unusual atmospheric phenomenon whereby an observer looking in the direction of the Sun can see, in addition to the main disc, four "sundogs" (parhelia) spaced regularly around a halo which circles the Sun; "spokes" join these images to the Sun. Very rarely, one can see also a second halo, with its own parhelia. The details of this model are too complicated to enter into here, but the explanation seems a lot more plausible than the invocation of ancient ufonauts.

Hopping forward in time by a good few centuries, we come to an event which occurred in 1290 at Byland or Ampleforth Abbey (depending on source):

> . . . when Henry the Abbot was about to say grace, John, one of the brethren, came in and said there was a great portent outside. Then

> they all ran out and lo, a large round silver thing like a disc flew slowly over them and excited the greatest terror. Whereat Henry the Abbot immediately cried that Wilfred was an adulterer . . .

Unfortunately, the original Latin version of this "contemporary account" was concocted by a couple of schoolboys, who included it in a letter to the London *Times* in 1953. This has been known for years, but the story still turns up in the literature – as does that of Cedric Allingham, who met a Martian in the north of Scotland. The photographs he took are blurred, but not so indistinct that one cannot see that the Martian is wearing braces, and that from the top of the UFO projects something which looks suspiciously like the flex of a hanging electric light. There is good reason to believe that "Allingham's" *Flying Saucer From Mars* (1955) was in fact written by a well known astronomer; "Allingham" disappeared from sight before publication. Another UFO hoax book, "William Robert Loosley's" *An Account of a Meeting With Denizens of Another World, 1871* (1979) by David Langford (b1953), has also made its way into the UFO history books.

One particular case seems to throw very grave doubt upon the alien-spacecraft hypothesis. Over the winter and spring of 1896–7 North America was bewildered by reports of airships. At that time, the airship was regarded in much the same way that we regard shuttle-type spaceships today. The mysterious airship more or less crossed the USA from west to east, taking five months over the journey. It was "seen" by countless people, some of whom were certainly hoaxers; hoaxers entered the picture more directly by carrying out pranks such as sending up hot-air balloons, while a number of people came forward to claim personal acquaintanceship with the airship's inventors, giving details as to its design and destination. The considerate crew of the craft even dropped letters overboard, one of them addressed to Thomas Alva Edison (1847–1931). In fact, it was Edison's public denunciation of the whole affair which brought the case to an end: sightings abruptly stopped.

There was a further airship craze, this time shorter-lived and largely confined to the northeastern part of the US, in 1909–10, this time the product of a hoax that got out of hand.

The parallels with modern UFO crazes need hardly be pointed out. Mistaken identifications of aerial phenomena were blown up out of all proportion by word-of-mouth retelling. From then on, whenever there was a "strange light" in the sky, the public was able, as its eyes strained to make out detail, to see an airship. On occasion it was even able to make out the silhouettes of the crew against the lighted windows! Exactly the same sort of thing happened in earlier centuries during the "witch craze": there were countless reports of witches being seen as they flew by on their broomsticks.

Of similar ilk was the "black helicopter" flap in northern England during the autumn and winter of 1973–4. The first helicopter sighting of note occurred on the night of September 18 1973 near Buxton, Derbyshire; more importantly, it was near a complex of quarries where explosives were stored. The police interest was not trivial. This was at the height of the IRA bombing campaign on mainland Britain, so the idea of the terrorists using a helicopter to raid explosives dumps was a frightening one. Indeed, over the whole duration of the flap, most of the sightings reported were made by police officers. There were some logistical difficulties with the scenario, however, and these soon began to surface within the media and in the hierarchies of the police. First was that most reports of the black helicopters indicated they were flying at an extremely low altitude, typically 15–30m. Helicopter pilots feel themselves pretty skilled if they fly at 150m during the daytime, so an altitude one-fifth or one-tenth of this at night would seem to require an impossibly proficient pilot. Second, helicopters are exceptionally noisy craft – think of the din if one flew over your house at a height of just 30m – yet most of the observers of the "black helicopters" either didn't mention the racket or commented on how quiet the vehicles were. (In one or two instances observers did mention the noise.) Third, and most important, helicopters guzzle high-octane fuel at a prodigious rate: where was it all coming from? Nobody could find any trace of the black helicopters ever refuelling.

Once these practical objections became dominant in the media coverage of the affair, the sightings started to drop off rapidly, and by about the end of March 1974 they were a thing of the past. (There have been minor recurrences since.) Again, this is a UFO flap in all but the identification of the aerial objects.

Most tales of UFO contacts are clearly delusions or inventions. But some won't go away that easily. The most famous is the case of Betty (1919–2004) and Barney (1921–1969) Hill, popularized in John Fuller's bestseller *The Interrupted Journey* (1966). In 1961 the Hills were driving home when they saw a light in the sky. They thought it might be a spaceship. When they reached home they discovered they had somehow "lost" a couple of hours. Then Betty began having dreams about being examined aboard a spacecraft by alien beings. Later, under hypnosis, she told of exactly this having happened during the Hills' "lost" two hours. Many years later, she produced a star-map drawn from memory: she said she could recall the aliens showing her something like it. It turned out to fit a hypothesis that the aliens came from the star Eta Reticuli, 37 light years away.

What appears to have happened is that the Hills were indeed frightened by a strange light in the night sky. Having read of UFOs, Betty, impressed by her period of fright, began to have imaginative dreams. She told her husband of these, and he was half-persuaded the dreams were rooted in memory. The pair underwent hypnosis with Bostonian psychiatrist Benjamin Simon, and more details came out. A word of caution: Contrary to popular belief, hypnosis does not lead subjects necessarily to tell the truth or unburden themselves of memories repressed into the unconscious; indeed, the hypnotic state encourages fantasizing, because it stimulates the imagination while at the same time relaxing any inhibitions the mind might have concerning the need for realism. Thereafter the subject can accept the induced fantasy as a genuine memory. All the same comments can be made about the dream state. Under hypnosis, then, Betty described not reality but what to her had *become* reality – *what she now sincerely believed had happened*. Barney did likewise but, not having had the dreams himself; was able to give only vague descriptions. Simon himself discarded the alien-spacecraft hypothesis completely: he was certain the experience had never occurred outside the Hills' minds – more particularly, Betty's mind.

The research behind Harvard psychologist Susan Clancy's book *Abducted: How People Come to Believe They Were Kidnapped by Aliens* (2005) was born from her interest in false memories. The subject is an interesting one, because – and this has to be emphasized – so far as the individual who has them is concerned, false memories are every bit as "real" as genuine ones; indeed, it's increasingly believed that what we call "memory" is itself a structure composed primarily of falsified memories – unconsciously embroidered versions of genuine events. (This is why, for example, eye-witness evidence is so notoriously unreliable in criminal cases.) Clancy began by interviewing victims of sexual abuse, but hit the obvious problem that there was no way of telling if their memories were true or false, or midway between. It was then that the work of psychiatrist John Mack (1929–2004) came to her attention. He had used hypnosis with alien-abduction claimers in an attempt to recover more detailed memories; now Harvard Medical School, suspicious, was starting to re-examine his research techniques. Alien abductions seemed an ideal field to Clancy, in that she could be fairly certain that any memories which emerged were false.

She and colleague Richard McNally began interviewing abductees, and fairly soon realized that most if not all of them were recording incidents of sleep paralysis. This is a not especially uncommon condition in which you are, in effect, both asleep and awake at the same time. It is characterized by, often, the feeling of paralysis, and, also quite often, hypnagogic dreams – startlingly lifelike hallucinations. A further common feature of the abductees whom Clancy and McNally interviewed was that they had spiritual inclinations but did not follow any conventional religious doctrine. Particularly vivid false abduction memories, Clancy reported, came from interviewees who had at some earlier stage undergone hypnosis in an attempt to "recover" their experiences. The conclusion of the two researchers was, Clancy wrote, that abductees' false memories came from "a blend of fantasy-proneness, memory distortion, culturally available scripts, sleep hallucinations, and scientific illiteracy".

Returning to the Hills, the evidence of the star-map might seem impressive. However, for statistical reasons, it would be highly surprising if a map of the kind produced by Betty did *not* "fit" with some or other pattern of nearby stars. In *The UFO Enigma* Menzel and Taves tell how they used a random-number generator to create star-maps, and found plenty of "good fits".

After Barney's death in 1969, Betty continued to be an avid UFO-watcher, going several times a week to a place near her home where she might observe as many as 80 UFOs in a single session; although the UFOs would blink their lights in friendly fashion when she called out to them, she had no further close encounters of the third kind. One night a pair of cops arrived and pointed out to her that the lights she was seeing were aeroplane lights. Not so, replied Hill: UFOs were perfectly capable of disguising themselves as aeroplanes.

The classic account of a UFO encounter concerns Captain Thomas Mantell (1922–1948), who lost his life as a result of it. In early 1948 a large glowing object was seen in the skies over Kentucky, and three USAF pilots who were already in the air were sent to investigate. Two returned to base but the third, Mantell, pursued the object, which he described over the radio as looking "metallic and tremendous in size". The object rose to a great height – over 6000m – and Mantell followed; soon radio transmissions ceased. Hours later the wreckage of his plane was found.

It seems likely Mantell lost his life investigating a balloon belonging to the US Navy. At the time these high-altitude balloons, which did indeed resemble large metallic objects, were secret. Mantell probably died because he flew too high for a plane which carried no oxygen equipment: he simply lost consciousness, and plunged.

A linchpin hypothesis of ufology is that the USAF knows what flying saucers really are, but isn't telling. In *Cults of Unreason* (1973) Christopher Evans cited Captain Edward J. Ruppelt (1922–1960), one-time director of two of the Air Force inquiries, projects Sign and Grudge. According to Ruppelt, in the early days after Kenneth Arnold's sighting the military took UFOs very seriously indeed, in the belief that they might – just might – be examples of a new type of aircraft being tested by the Russians. However, the research carried out was fairly haphazard, and was confused by erroneous and self-contradictory sightings. Thus specific queries from the press were received with blank looks and general evasion – because no one really knew what, if anything, they were talking about. From this situation sprang the great USAF "cover-up".

Perhaps UFOs are "survivals" of aircraft built by the ancient Indians – a suggestion put forward by Brad Steiger. He tells of G.R. Josyer's translation of the *Vymankia Shastra*, an ancient Sanskrit text which deals with the art of flying around in *vimanas*, which seem at first sight to have been some sort of fighting aircraft. But aircraft with a difference. They appear to have been powered mentally, rather than by engines, and to have had various little extras: good pilots could make their machines zigzag across the sky, become invisible, look like a cloud, or even change shape. Little wonder that Josyer feels the *Vymankia Shastra* could transform the science of aeronautics and that its translation by him is one of the two truly historic events of the 20th century – the other being the bringing back to Earth of Moon rock. Most useful of all must have been the ability to make one's *vimana* appear like a "heavenly damsel bedecked with flowers and jewels".

As Steiger points out, UFOs are well known to zigzag

across the sky or abruptly disappear, and he adds: "Likewise, the transformation of UFOs to cloud has been reported by many observers." And cloud to UFOs, no doubt. As for that gorgeous damsel? – why, there are those appearances in the sky at Lourdes and elsewhere of female forms generally believed to be of the Virgin Mary.

Talk of cloud-like UFOs leads us inevitably to orgonomy. Orgonomy, which developed into a more or less complete knowledge system, was the brainchild of Wilhelm Reich (1897–1957), an unorthodox psychoanalyst and one-time disciple of Freud. Reich had earlier proposed that mental illness can cause a tightening of the muscles: massage the muscles and the mental illness will go away. This "vegetotherapy" may well be beneficial, especially in cases of chronic depression.

Reich's discovery of the fundamental motive force of the Universe, orgone, was based upon his belief that neuroses were the result of the individual's inability to achieve a satisfactory orgasm. He argued his case in *The Function of the Orgasm* (1927) and, while the idea was hardly approved by his fellow psychoanalysts, it was not thought ridiculous. He considered the "afterglow" of a good orgasm: perhaps the body was permeated by something he called orgone energy, which during orgasm was concentrated in the genitals; afterwards it gradually dispersed itself throughout the body once more. Thus sex fulfilled roughly the same function as does the heart in the blood-circulatory system, ensuring that orgone energy was kept pumping around the body. (Reich's views on sex are of interest not only because of his work on the orgasm. He believed homosexuality to be a bizarre evil, and hated all homosexuals for that reason. And he advocated promiscuity for himself while being almost insanely jealous of his long-suffering wife.)

Orgone energy was to be found not only in the bodies of sexed organisms but throughout the Universe, its main source being the Sun and stars. Thus, if the body was suffering from a lack of orgone, the thing to do was to sit outside for a while

Cave painting from Jabbaren in Algeria.

under the open sky (orgone was blue: the sky was obviously full of orgone). However, this treatment was only a very mild one, and so Reich sought to find a way of somehow *concentrating* natural orgone energy. Thus was born his famous orgone box – or, more properly, the Orgone Accumulator. This consisted essentially of a box large enough to hold a seated human being, made of an inner layer of metal surrounded by an outer layer of wood or other organic material. The wood, being organic, attracted orgone from the atmosphere and transmitted it through to the metal; the metal, however, being inorganic, could not retain the orgone energy, and radiated it inwards to the box's lucky occupant. Moreover, the amount of orgone inside the box would build up – rather as if the inner surface of the metal acted like the silvered inner sides of a vacuum flask. Orgone could come in but was unable to exit because simply reflected off the metal walls.

Reich made some unusual claims for orgone. It could be seen under the microscope and detected by various other everyday scientific instruments, such as the thermometer and the Geiger counter. Also, you could see the elementary particles formed when two beams of orgone energy met: these "bions" could be seen either through the microscope or in the form of spots before the eyes.

In later years Reich came to believe fervently in UFOs, which he thought were sent to attack him, personally – by this time he was convinced of his own messianic rôle. However, he knew exactly how to retaliate. Clouds, he had earlier realized, were manifestations of destructive orgone energy, and he had invented a device called a "cloudbuster" to disrupt them. This comprised a collection of hollow tubes connected at one end to running water; the destructive orgone energy was drawn down the tubes and dissipated in the water. The same device could be used, he reasoned, to combat UFOs – for their only possible means of propulsion had to be that ultimate power-source, the orgone drive. And from an orgone drive you get dangerous exhaust fumes of – yes! – destructive orgone energy. Quite why dispersing their exhausts should unduly perturb the hostile saucer-pilots is an enigma, but it all made sense to Reich. While a vast cosmic war raged between friendly and hostile UFOs – and who knows how he told friend from foe – Reich single-handedly fought off the menace from space and saved the world.

Various ufonauts warn us against the dangers of nuclear war – and one cannot but laud them for so doing. However, it would be more reassuring if some of them didn't seem to think the energy of a nuclear explosion comes from the splitting of the hydrogen atom.

Photographs of UFOs – even films/video – abound. To the converted these constitute the final proof. However, the depressing truth is that almost all of these photographs can be shown to be not of spacecraft. Some are hoaxes, others are genuine photographs of meteorological phenomena or distant aircraft, and yet other images are due to faulty film, faulty processing or faulty photography. It is not simply a question of being able to explain away these photographs without invoking flying saucers: it is the plain fact that they *do not show* flying saucers. Of the small remainder, virtually all do indeed show something unidentifiable, but this is due to some deficiency or other in the photograph: the object could be anything. This leaves a few photos and videos that genuinely puzzle. Once again, the leap from this to the confident assertion that the images are of alien spacecraft is a leap too far.

The much maligned Condon Report (1968) stated that, out of 7641 UFO sightings, 1566 could not be identified as either having perfectly natural causes or being hoaxes. This is about 20%, which is a startlingly high proportion until you discover that 1313 were unidentifiable simply because there wasn't enough reported information to make even a guess. Still, that leaves 253 genuinely inexplicable sightings – only 3.5%, but still significant. Until you consider that some of those could be accounted for by, say, undetected hoaxes, wrong information (again, remember the unreliability of *all* eye-witness accounts), or, most interestingly, phenomena as yet unknown to science. Remember also that the investigators were working always with secondhand data: they couldn't go out and repeat the observations themselves. Viewed in this light, it is perhaps surprising so few of the reports were inexplicable.

Nevertheless, a hard core of ufologists claims those 253 sightings necessarily represent genuine observations of alien spaceships. The logical flaw here is obvious.

UFO RELIGIONS

In Spring 1954 a Londoner, George King (1919–1997), had something of a surprise. Suddenly a voice spoke to him, from out of nowhere, saying: "Prepare yourself. You are to become the voice of Interplanetary Parliament." (He was later to discover that Interplanetary Parliament is an august governing body that holds regular sessions on Saturn.) About a week after this first experience, an oriental yogi entered King's bedsitter – through a locked door – to confirm that he had indeed been selected to act as a representative of the Cosmic Masters, and proceeded to teach him how to establish contact with these entities. And so, some months later, King found himself in direct contact with a Master codenamed Aetherius who dwells upon Venus.

Early in 1955 King held a first meeting, in London's Caxton Hall, to promote the discipleship of Aetherius: he was able to go into a trance and relay directly to his audience the very words of the Master. The founding of the Aetherius Society followed not long after, in 1956 (a US branch was established in 1960). Thereafter King and his adherents cooperated with Aetherius and the other Cosmic Masters in various bizarre exercises whose purpose has been to protect Earth from all sorts of nasty fates. Those crises can be seen as the diagnostic symptoms betraying what the Aetherius Society really is, behind its disguise of interplanetary jargon: they occur with reasonable frequency, thereby binding adherents with the repeated threat that a relapse of faith might have dire consequences; they are, after much striving, headed off through the invocation of the Cosmic Masters, so that, to any sceptical inquiry as to why it is that none of the dreadful threats have ever come to anything, it can be replied that this is a measure of how effective the workings of the Cosmic Masters truly are. In short, these are the hallmarks of a pseudoreligion – all the hallmarks bar one: it seems the Aetherians, to their credit, are not much interested in exploiting the credulity of their neophytes for financial gain.

In due course King became Sir George King, by courtesy

of a knighthood conferred upon him by the Sovereign and Military Order of John of Jerusalem and the Knights of Malta. He acted as the mouthpiece for various other cosmic voices: regular favourites were Jesus Christ, St Peter, the Buddha and a Martian scientist called Mars Sector Six. It was Mars Sector Six who, through King, gave the society its slogan: "Service is the Jewel in the Rock of Attainment."

It is the contention of the Aetherius Society that UFOs are craft from the other planets of the Solar System, notably Venus and Mars. While those planets are utopian and their inhabitants – as represented by the Cosmic Masters – are wise and good, the Earth is still teetering on a knife-edge, pulled this way and that by the twin but opposed Forces of Good and Forces of Evil, the war between which is raging not just in the vicinity of the Solar System but throughout the cosmos. The intention of the Cosmic Masters is to pour the yogic vital force prana down onto the Earth to help us stay on the side of the Forces of Good. Members of the Aetherius Society can assist in this work through the medium of communal "spiritual pushes" and Prayer Power. According to this latter notion, prayer can be used to generate spiritual energy which can then be stored in a physical container; a bottle will do as well as any, but objects as large as mountains can be used as receptacles. Later the spiritual energy can be released in concentrated form either to avert disasters or at least to reduce their effects. And these need not be just run-of-the-mill catastrophes such as earthquakes: the combined efforts of the Aetherius Society and the Cosmic Masters have several times saved us from holocaust – including, in the early 1970s, a potentially ruinous attack of Pole Shift.

One of the most influential UFO religions on an international scale, the partially ufological Scientology aside, is Raëlianism, which currently has a claimed membership approaching 50,000 worldwide (some reckon the true membership is about half this figure). The International Raëlian Movement, as it is more formally called, was born in 1973 when journalist Claude

Vorilhon (b1946), now known as Raël, felt an impulse to skip work at his office in Clermont-Ferrand, France, and instead visit the nearby extinct volcano of Puy-de-Lassolas, where he met a green-skinned humanoid from a flying saucer. The alien, who was 1.5m tall, briefed him on this and the five succeeding mornings as to what he was to do, which was tell the world The Truth: large portions of the Bible have been severely misinterpreted, most notably the use in *Genesis* of the word "Elohim", which means not "God" but "those who came from the sky" – in other words, space beings: it was the Elohim who created life on Earth, not God, and all the world's major prophets have been messengers of the Elohim, not of God, with Raël the latest in that prophetic line and, for some reason, the first to get the story straight. The prophets, Raël included, have also all been Nelphelin, the product of mating between an Elohim father and a human mother, as mentioned in *Genesis*. The Elohim, the alien explained, are eager to return to Earth and share their advanced technology with us, but this they will not do until an embassy has been built for them in Jerusalem and until there is world peace. They require this embassy because, if they landed elsewhere on the globe, they might be seen as favouring one particular non-Raëlian religion or ideology over another.

Raël moved swiftly to tell the world, as instructed, self-publishing the first of several books, *Le Livre que Dit la Vérité* ("The Book that Tells the Truth"), in 1974. He also moved to raise the money to build the Jerusalem embassy, but hit a problem that has yet to be resolved: the Israeli authorities are reluctant to permit the embassy to be built there. In part this may be because the Raëlian Movement's original symbol was a Swastika with a Star of David; this was changed in 1991 to show a Star of David with a spiral galaxy, but still the Israelis are resistant.

The Elohim originate on a distant planet. Long ago, their civilization developed genetic and microbiological technology to a high level, and they were able to create life from DNA. They searched for a different planet upon which they could conduct their genetic experiments in isolation from their homeworld, and discovered the young Earth. They built labo-

Australian rock paintings could show spacemen or Vondjina, the Creation god.

ratories in what is now Israel and Palestine, and in this "Garden of Eden" they created plants, animals and finally humans. Initially the humans were allowed to live in the laboratory compound, but they proved destructively aggressive and so were thrown out. Although from time to time since then the Elohim have mated with humans to produce prophets, for the most part we've been left to fend for ourselves, with the Elohim merely passively watching. Now, however, we have entered the Age of Apocalypse, whose start was signified by various events such as the dropping of the atom bomb, the creation of the state of Israel, and the development of global telecommunications (a time when we can "send our voices across the oceans"). Soon the final jigsaw piece of the Age of Apocalypse will be set in place, when advances in genetics enable us to create life from inert matter. We will at last be sufficiently advanced that we can properly understand our origins.

According to the Raëlians, God is not an individual but an infinite and infinitely diffuse entity – a sort of universal essence. There is no afterlife: the soul is extinguished with the life of the individual; however, there is the possibility of eternal life through the regeneration of the individual from her or his DNA. Life and reality exist on infinite levels: the Earth is but a single atom of a huge being who lives on an appropriately huge planet, which itself is merely an atom of an even huger being, and so on, with the same principle extending also in reverse, towards tinier and tinier worlds and their inhabitants.

Raëlians believe very strongly in the merits of human cloning, especially as a way in which infertile and gay couples may have children that are genuinely their own. To this end, in 1997 the Raëlian Movement founded the company Valian Venture Ltd, which conducts research on human cloning and, through its project called Clonaid, a service to help couples who wish to have cloned offspring and have the necessary $200,000 fee. Another service, whose title is self-explanatory, is Clonapet. In the late 1990s the Raëlian Movement made a determined effort to hire Dr Richard Seed, part of the team responsible for creating the cloned sheep Dolly. In December 2002 Clonaid's CEO Brigitte Boisselier claimed the company had cloned a human child, a girl-baby called Eve; in January 2003 Boisselier claimed a second cloned child had been born and thereafter that several others would soon follow in different parts of the world. These announcements naturally created international headlines, but further details (and evidence such as DNA samples) have not been forthcoming, the excuse being offered that Boisselier might find herself in jail for having violated various countries' anti-human-cloning laws. The focus of the Raëlian Movement on cloning is not merely an opportunistic one: it is a religious duty, because according to their principles it is only through cloning that humans can attain eternal life and thus become one with the Elohim.

In 1975 Marshall Herff Applewhite (1931–1997) and Bonnie Lu Truesdale Nettles (1927–1985), also known as respectively Bo and Peep, or from about 1976 as Do and Ti (after the musical notes), or collectively as The Two, announced they had come to Earth by UFO from the "level above human", to which level they would return within a few months, taking along with them any lucky people who were prepared to join "the Process". Oh yes, and if you wanted to join up you had to abandon all your worldly goods and your previous life. Because of the requirement to abandon worldly goods and existing relationships, including one's family, the ufological religion started by

The Two, Heaven's Gate, later renamed Total Overcomers Anonymous, remained small, numbering at its largest perhaps a few hundred adherents; a further deterrent to new recruiting may have been that, as months passed and became years, there was still no sign of the promised ascent. The death of Ti (Truesdale) in 1985 of liver cancer may also have caused some disciples to question their leaders' extraterrestrial origins and immortality.

Everything changed with a vengeance in March 1997. By this time Applewhite was living in a multimillion-dollar mansion in a suburb of San Diego, California. It was in this mansion that he and 38 Heaven's Gate disciples were found dead: they had committed communal suicide by swallowing a cocktail of barbiturates and vodka. Alternatively, they had not killed themselves but, rather, departed their physical bodies in order to ascend to a higher level. They were encouraged in this by the arrival in the skies of the spectacular Comet Hale–Bopp, which Applewhite had claimed as the long-delayed signal from the extraterrestrials that the time had come. A further inducement was an urban legend that a spacecraft had been spotted travelling in the comet's tail; this craft, Applewhite seems to have believed, was bearing Truesdale back to him and their shared disciples. In a video Applewhite made before the mass suicide, he stated: "We do in all honesty hate this world." More prosaically, he and Truesdale had created a reality in which they felt comfortable, a reality that had little to do with the one most of the rest of us know, and had then succeeded in essentially brainwashing others to accept this artificial reality. As real and artificial realities increasingly clashed, the more and more untenable the artificial version became. The options for devotees then became either to shake themselves out of the mindset and leave the group or to retreat precipitously from the position of having to accept that they'd thrown away everything – including much of their lifetimes – for the sake of a complete nonsense. Thirty-eight devotees, plus Applewhite, chose that latter course.

Alchemists at work. From *The Art of Distillation* by John French, 1651.

CHAPTER 5

HARD SCIENCE

IN 1962 THE RUSSIAN CHEMIST N.N. Fedyakin announced the discovery of what appeared to be a new form of water, which had a lower freezing temperature, was more viscous (i.e., less runny), and was denser. This kind of water could be found when ordinary water vapour had condensed in very narrow glass or fused-quartz tubes; it was christened "anomalous water", "orthowater" or "polywater" – the last because it was assumed to be a hitherto unknown polymer of water.

More recent research has revealed that polywater is merely ordinary water contaminated by ions which it has dissolved from the tube walls – and, it has been suggested, from the skin of the experimenters' fingers.

The term "hard science" does not refer to the science people find hard in school, although for many there's a fair level of crossover between the two descriptions. The hard sciences are those like physics and chemistry where the scientist is dealing with physical material, as opposed to the soft sciences, like psychology. It's hard science that, among other things, makes the wheels of our technology go round.

The biological sciences are mainly discussed in the next chapter: in the coming centuries, with the direction much of our technology is currently taking, biological sciences such as genetics may prove to be the most important hard sciences of all . . . which is odd, because only a few decades ago biology and medicine were often thought of more as soft sciences.

ALCHEMY

To understand alchemy one must understand the idea of the Aristotelian elements. To do so it is important to rid oneself of the idea that matter might be made of atoms. Matter is instead composed of *qualities*, and is itself only one of three components which go to make up a lump of (say) lead. The other two components are form and spirit. Thus a lump of lead is made up of the same sort of matter as a lump of gold, but has different "form" – i.e., properties. Neither lump contains much spirit – not as much as air, say, and certainly not as much as God, who is purely spiritual.

Aristotle (384–322BC) felt there was only one kind of matter, which could take many forms. The fundamental forms, the elements, were four in number: earth, air, fire and water. The qualities which these gave to objects or aggregations were: earth, cold and dry; fire, dry and hot; air, hot and wet; water, wet and cold. This mode of thinking led directly to the belief in the four humours (see page 265); and it was from the base of the Aristotelian elements, and the intuitive logic behind them, that the alchemists started their quest.

Because these four were only fundamental forms of the single type of matter, and therefore not "elements" in any sense of the word we might recognize, they could be transmuted into each other. This gave rise to Aristotle's elegant vision of what we would today call the "hydrologic cycle". Briefly: The Sun's heat changes water into air (throughout we are talking of the elements, not the chemical compounds). Heat rises, so the heat in this air pulls the air up to the skies. The heat then leaves the vapour, which thus becomes progressively more watery again, and this process is marked by the formation of a cloud. There is a positive feedback: the more watery the mixture in the cloud, the more the water (naturally) drives away its opposite, the heat. Thus the cloud gets colder and, overall, shrinks. This contraction restores true "wateriness" to the water, which falls as rain or, if the cloud is by now so cold that it has frozen, hail or snow.

It is startling that, in Aristotle's model of the hydrologic cycle, all the wrong reasons produced something alarmingly close to the right answer.

The science of alchemy was probably born about the time of Christ in Graeco-Roman Egypt. There was a separate oriental alchemical tradition, but the relationship and links between it and the occidental tradition are problematic. The mainstream of alchemy petered out around the time that the new, quantitative, rationalist methods of the physicists were filtering through to chemistry, although there are still a few practitioners today.

Understanding alchemical ideas is hindered in several ways. First, the simplistic popular view that the quest of the alchemist was merely to isolate the Philosophers' Stone and use it cupiditously to transform base metals into gold shrouds a genuine quest for mental and spiritual advance. Second, the alchemists took great pleasure in making their own writings as incomprehensible as possible: instant knowledge was not to be available to the uninitiated. And, third, the scheme of Nature that figured in the alchemists' conception is totally divorced from anything in our modern worldview.

To the alchemists, "spirit" and matter could be – and were – mingled. Some things contained more spirit than others – and God was, of course, the purely spiritual stuff. The ability to use "spirit" in their alchemical experiments was important – and, naturally, very difficult – but it was vital if they were to succeed. Imagine you wish to transform copper into gold. Heating copper with sulphur will reduce it to a black mass, which is obviously the "basic stuff" of copper, its metallic "form" having been ousted by the treatment. (In fact, the black mass is copper sulphide.) So far, so good: now for the more difficult second part of the experiment, the introduction of the "form of gold" to this mass. To do this you have to be able to manipulate and admix suitable quantities of "spirit". It was this complicated second part of the problem that stymied the alchemists for over a millennium and a half.

Coupled with what one could rather uneasily term "practical chemistry" like this was the theory that, just as transforming base metals to gold was making matter "more perfect", so success in the quest would make the alchemist himself * "more perfect". It is no coincidence that the Philosophers' Stone was thought to be the elixir of life, conferring physical and spiritual immortality.

The interest of the aristocracy in alchemy, however, was focused less on the spiritual voyage, more on the transmutation of base metals into gold and the manifold increase thereby of their wealth. The gate was wide open for fraudulent alchemists to prey upon such avarice. The Hapsburgs seem to have been particularly gullible in this respect. In 1658 Ferdinand III (1608–1658) himself witnessed the creation of a nugget of gold by the alchemist J.K. Richthausen. The Czech alchemist Wenzel Seyler turned silver medallions into gold for Leopold I (1640–1705), and was accordingly ennobled by the Emperor. And so on. Maria Theresa (1717–1780) attempted to put an end to such nonsense by banning all transmutation attempts from her realm, but after she died the old obsession returned to the Hapsburg court. As late as the 1860s two self-styled alchemists were, before sense prevailed, well on their way to duping the Hapsburgs out of substantial sums with the usual promises of turning silver into gold.

While it is a cliché to say that alchemy fathered chemistry in the same way that astrology fathered astronomy, it is pertinent to note that alchemy later gave a tremendous boost to the nascent discipline of the Earth sciences, too: when this had advanced to the point where (say) analysis of minerals was desirable, already on hand were the techniques of alchemy. Similarly, while alchemy gave rise to some bizarre ideas about the Earth, it did at least focus attention on such phenomena as earthquakes and volcanoes. Alchemy, in sum, provided a fertile ground in which could be planted the seeds of much of our modern science.

* Or *her*self – there were a few female alchemists, most notable among them being a shadowy figure called Mary the Jewess, who possibly lived in the 3rd century AD.

One of countless supposed portraits of Nostradamus done long after his death.

ASTROLOGY

Astrology originated in Babylonia, probably two or three millennia BC, and it did so in difficult observational circumstances. In theory in that part of the world the observing conditions should have been excellent: the skies were clear, the weather was good and the terrain flat. In practice, however, wind-blown dustclouds often obscured the horizon. Were astrology zenith-oriented, then it would be easier to dismiss the origin of astrology as being perhaps just a recreation that got out of hand. But astrology is very *horizon*-oriented: to the astrologer, the region of the horizon is the most important part of the sky. Manifestly, then, our ancestors thought the dispositions of the planets among the stars really was important to us below, or they wouldn't have put so much effort into carefully observing their rising times.

That there is some truth in astrology was supposedly shown by Michel Gauquelin (1928–1991), a French statistician with apparently unlimited reserves of doggedness. He showed there was a strong correlation between *distinguished and successful* people and the planet rising at the time of their birth. If Mars or Saturn (or both) is rising as your child is born, there is a strong chance he or she will grow up to be a doctor or scientist; if Mars or Jupiter, the child may become a soldier or athlete (note that Mars "creates" both soldiers and scientists). In this scheme there is, of course, no place for the astrologers' other staple, the Zodiac. Gauquelin's results created quite a stir when he announced them in *The Cosmic Clocks* (1967), but later re-analyses failed to show the same correlations. Nevertheless, devotees of astrology are still fond of citing his experiment to support their case. Curiously, they rarely mention *another* experiment Gauquelin performed, in which he sent the accurately compiled astrological birth-chart of the serial killer Michel Petiot, executed in 1946, to a selection of 150 people, saying to each that this was their own birth-chart, and asked them for the comments of themselves and their friends/families. A staggering 94% of the recipients said the birth-chart described them excellently, and of these almost all reported that their friends/families concurred. The birth-chart of the murderer, by the way, included such comments as "endowed with a moral sense which is comforting", so can hardly be regarded as a successful piece of analysis of the personality upon whose astrological details it was based.

Is there "something in" astrology? Certainly the outpourings of newspaper astrologers and indeed of "professional" astrologers are valueless, and bring with them untold quantities of pseudoscientific baggage. To cite just one example, Alfred Witte (1878–1941), who founded the astrological group known as the Hamburg School, was seemingly so inspired by the talk of astronomers that there must be a planet beyond Neptune that he went ahead and found eight of them – although, surprisingly, not Pluto, which had to wait until Clyde Tombaugh (1906–1997) located it in 1930. Astronomers have yet to confirm Witte's epoch-making discoveries, yet the members of the Hamburg School incorporate all eight when drawing up horoscopes.

It is commonplace that astrology fathered astronomy. But its influence was greater than that. Franz Mesmer (1734–1815) conceived invisible force-fields operating between distant bodies in the Universe in his attempts to explain why astrology worked. Later he was to be more definite about all this, producing the important concept of animal magnetism (see page 315).

PHLOGISTON

In 1700 the German physician Georg Ernst Stahl (1660–1734) invoked "phlogiston" to explain what happens when things burn or rust (he had realized the two processes were essentially the same). He suggested that a burning substance was losing an elementary and undetectable principle. Unfortunately, while this neatly explained why the original substance (rich in phlogiston) was heavier than its ashes (deficient in phlogiston), it far from explained why a rust (deficient in phlogiston) was actually heavier than the original metal (rich in phlogiston). While we today would raise our eyebrows at a substance which in some but not all cases had "negative weight", in Stahl's time physical chemistry – like its predecessor, alchemy – was not really concerned with such trifles as mass. The fact that air was required for both burning and rusting was simply explained: the air was needed to transport the phlogiston away from the substance.

The phlogiston theory was of considerable importance during the first three-quarters of the 18th century, since the processes occurring during combustion were then of especial interest – it was the dawn of the age of steam.

It is ironic that the English chemist Joseph Priestley (1733–1804), a devout supporter of the phlogiston theory, contributed to its downfall. He heated mercury in air to form red mercuric oxide (of course, he didn't realize what the substance actually was). He then applied concentrated heat to the oxide and noticed that it decomposed again to mercury while giving off a strange gas in which things burnt very vigorously and brightly; clearly, he concluded, this gas must be phlogiston-poor. A little earlier, the Scottish physician Daniel Rutherford (1749–1819) had found that, if you kept a mouse in an enclosed space until it died of suffocation, then burnt something in the air until it would burn no more, you were left with a form of air in which creatures could not breathe and substances could not burn. Clearly this air was so rich in phlogiston that it could accept no more. Rutherford had called it "phlogisticated air", and so Priestley called his own new gas "dephlogisticated air".

In 1774 Priestley visited the French chemist Antoine Lavoisier (1743–1794) in Paris and told him of his experiments. These Lavoisier repeated, and he swiftly realized the truth of the matter: air was made up of a mixture of two gases, one of which supported combustion and one of which would not. It was clear that in both burning and rusting the substance was picking up one of these gases from the air; Lavoisier called it "*oxygène*", meaning "acidifying principle", because he thought it was present in all acids (he was wrong). The other gas, Rutherford's "dephlogisticated air", Lavoisier called "azote" ("without life") – the name was later changed to "nitrogen" ("giving birth to nitre").

A far more elegant theory of combustion dates back as far as the 2nd century, when Philon of Byzantium probably inferred that, in terms of the Aristotelian elements, combustion turned air particles into fire particles, which were smaller, He had noticed that, if you burn a candle in an upturned container (e.g., a tumbler), the open end of which is submerged in water, when the candle goes out the water has risen a little inside the container. He suggested, correctly, that this was because some of the air had been used up in the combustion; but he proposed that what had happened was that this portion of the air had been converted into the smaller fire particles, which were able to escape through pores in the container's walls.

CALORIC

Rather like phlogiston, caloric was a weightless fluid, a quality, which could be transmitted from one substance to another, so that the first warmed the second up. Today we realize that what is in fact being transmitted is not caloric but heat energy.

All substances contained caloric – the trick was to get at it. Two pieces of wood rubbed together would give heat because little bits of the wood were being abraded off, releasing the caloric trapped within. Where a kettle was being heated over the fire, the fuel gave up its caloric to the flame, which passed it on to the metal, which passed it on to the water.

Benjamin Thompson, Count Rumford (1753–1814), scotched the theory towards the end of the 18th century. While overseeing the boring of cannon for the Elector of Bavaria, he noticed that a great deal of heat was being generated. According to the caloric theory, this was because the removal of shards of brass was releasing some caloric from the body of the cannon; but Rumford noticed that, if the tools were blunt and removed little or no metal, *more* heat was generated rather than less – i.e., what was happening was exactly the opposite of the theory's prediction.

He measured the amount of caloric obtained and discovered that, were this quantity to be poured back into the cannon, the metal would melt. Clearly, then, it was impossible for the cannon to have contained that much caloric in the first place. Rumford concluded that the heat content of an object is the measure of some sort of vibration within it – which vibration was, in the case of the cannon, being bolstered by the friction of the tools. In other words, he had recognized the relationship between heat energy and the physicist's concept of "work". You put work into the rubbing of two sticks and you get heat out.

Later Rumford carried out a further experiment in which he weighed some water as both liquid and ice, and found no detectable difference in weight. According to the caloric theory, ice contained less caloric than water and so the inescapable conclusion was that caloric, if it existed at all, must be weightless. A few decades earlier this would have been acceptable but, 25 years before, Lavoisier had shown that the weightless phlogiston was a myth, and so it seemed caloric, too, was ready to disappear from the textbooks. Nevertheless, half a century passed before, in 1849, James Joule (1818–1889) read a paper on his establishment of the "mechanical equivalent of heat" to the Royal Society.

The final death-knell for the caloric theory came a few years later with the work of James Clerk Maxwell (1831–1879) and the formal launching of the kinetic theory, still a cornerstone of physics. The kinetic theory has it that the heat content of a body is equivalent to the sum of the individual energies of motion (i.e., kinetic energies) of its constituent atoms and molecules.

LEVITY

To say that levity is the opposite of gravity is the literal truth, but may be misleading: there is no suggestion that levity could be equated with antigravity. It is obvious to all that objects are drawn downwards, towards the Earth, to varying degrees: feathers are "pulled" less than hammers. In earlier centuries it was reasonable to suggest that the amount by which an object was pulled downwards (or, in more sophisticated post-Newton times, resisted changes in its state of motion) was a measure of the amount of gravity the object contained. Gravity, then, was a principle much like, say, caloric.

But, if objects could contain gravity, surely also they could contain its converse, levity? The more levity there was in an object, the lighter it would be.

The matter assumed some importance towards the end of the great phlogiston affair. Stahl, the originator of the phlogiston theory, had made the connection between burning and rusting, showing that in both cases the original substance was "losing phlogiston"; but, while ashes are generally lighter than the original substance, rusty metals are heavier than unrusted ones. Clearly, later scientists proposed, some phlogiston possessed gravity (its departure made the substance lighter) while some possessed levity (its departure made the substance heavier).

GLOBULISM

During the 18th century it was a popular theory that most if not all forms of matter were filled with little globules – which were possibly atoms: John Dalton (1766–1844) was to formulate his Atomic Theory at the very beginning of the next century, in 1803. You could even see these little globules under the more powerful microscopes of the day.

Unfortunately, as soon as the achromatic lens was developed, the globules became undetectable under even the best instruments – indeed, *particularly* under the best instruments. They had been merely optical illusions generated by the simple lenses of the early microscopes.

CORPUSCULAR THEORY OF LIGHT

That a ray of light might be a stream of tiny particles (corpuscles) is an ancient notion. Oddly, it wasn't quite subscribed to by Democritus (c470–380BC), even though he believed *all* things were composed of atoms and of void (the higher the atom-to-void ratio, the denser the material). He had some difficulty in explaining how we are actually able to see things – it would be too unlikely to propose that all objects emitted streams of atoms which entered our eyes and then affected the configurations of the atoms therein. What he suggested was that, as you look at (say) this page, it is constantly "imprinting" the air with images of itself. The air travels towards your eyes, bringing the images with it.*

Disturbingly similar was the theory of substantialism propounded by the Reverend Alexander Wilford Hall in his

* With what we may regard with hindsight as considerably more success, Democritus proposed a very similar theory to explain how we hear sounds.

The Problem of Human Life (1877). All "forces" and "radiations" are made up of atoms – i.e., they are substances. It is true that if you smell a rose it is because particles from the rose are reaching your nose. Surely, then, if you hear a guitar it is because sound particles from the guitar are reaching your ears. Hall stated that forces like gravity and magnetism, and radiations like light, were made up of atoms much tinier than those which make up matter. This sounds uncannily like the ideas of modern physicists who posit, for example, the "gravity particle", or graviton. But the similarity is, alas, more apparent than real.

The corpuscular (or ballistic) theory of light had importance for more than a century because it was supported by Isaac Newton (1642–1727), and has returned to prominence again in the past few decades. The debate concerned whether light was made up of waves or of particles. In the light of Newton's science, it seemed unlikely that it was made up of waves: after all, if you shout at someone in the next room they are likely to hear you because the sound (which is a wave motion) can go round the corner of the door; yet light cannot do this. Therefore Newton believed light must be particulate in nature. (But he was not dogmatic about this: in *Opticks*, 1704, he showed an awareness that there were problems with the corpuscular theory.) Because of Newton's eminence his theory held sway for well over a century, until Thomas Young (1773–1829) in 1803 demonstrated interference.

The significance of Young's experiment is that, if a source of light is a point, the shadows it casts are sharply defined – or seem so; in other words, the light is not going round corners. But by Newton's time Francesco Grimaldi (1618–1663) had already demonstrated the existence of diffraction, a phenomenon in which light very definitely *does* bend at corners, only not very much. Newton himself observed diffraction (it seems he was not aware of Grimaldi's work), but was unable to explain it. Young's experiment proved beyond doubt that light has a wave nature. Think of two musical notes which are not quite in tune with each other: as they are played, you hear "beats" because the waves from the two instruments are not quite in step – that is, sometimes the waves reinforce each other and sometimes they negate each other. If light could be shown to do the same as sound, then light was a wave motion – because adding one *particle* of light to another will never give you a result of zero, or darkness. Young passed light through a pair of narrow parallel slits and found that, sure enough, the result was the projection of a barred image. In the interference pattern thus produced, the dark bands are the result of the light waves negating each other, the bright bands areas where the light waves are reinforcing each other.

Despite Young's demonstration, the debate did not die, as evidenced by books like R.A. Waldron's excellent *The Wave and Ballistic Theories of Light* (1974). The reason for the continuing discussion was that, in many ways, light behaves more like a string of particles than like a wave motion. Today it is realized that in truth light is *both*. The two views are not irreconcilable. The fundamental particle of light (and of other electromagnetic radiation) is the photon; yet here we are using the word "particle" in a sense rather different from anything Newton could have imagined.

STRANGE RAYS

In 1903 the prominent French physicist René-Prosper Blondlot (1849–1930) discovered N-rays, produced naturally by various materials, including many of the metals, but also by the human nervous system – notably, when people talked, by the part of the brain that controls speech, Broca's area. (Blondlot called them N-rays in honour of his employer, the University of Nancy.) His findings were confirmed by other French scientists, although outside France experimenters had difficulty in reproducing his results.

Using adapted spectroscopic equipment – in which the lenses and prisms were made of aluminium – Blondlot could project N-ray spectra; this had to be done in darkness. An observer at one such demonstration was the US physicist

Robert W. Wood (1868–1955). While Blondlot was describing the N-ray spectrum he was projecting, Wood quietly removed the prism from the N-ray "spectroscope". Blondlot continued his demonstration, unperturbed. In 1904 the French Academy of Sciences bestowed on him the Leconte Prize. In that same year, though, Wood published an article recounting this experience, and non-French scientists promptly stopped looking for the elusive N-rays. Within France, however, physicists not only kept looking for them but also in many cases finding them. Among these physicists were distinguished individuals such as André Broca (1863–1925) and Jean Becquerel (1878–1953).

It is clear that Blondlot was by no means a hoaxer; he sincerely believed he could see the N-ray spectra, and that he had made an important discovery. While this is understandable, it is difficult to know how all those other French scientists were able to duplicate his results. Was it simply the case that their French nationalism or their respect for Blondlot was leading them into self-deception? Or was it that the climate in French physics was one of acceptance of N-rays as proven entities, and so experimenters saw what they expected to see?

Of similar ilk were Shearer-rays. X-rays allow us to examine bones through the thin curtain of human flesh – but imagine the uses of a radiation which allowed us to do the same for bodily organs. Such radiation was discovered during WWI by a British Army orderly named Shearer, about whom nothing else is known. He also devised equipment with which to put the radiation to practical use. He was promptly promoted to the rank of captain, and encouraged to continue his researches.

But, as John Sladek (1937–2000) reports in *The New Apocrypha* (1974), a Shearergraph was not just a medical tool. When it was suggested to Shearer that he could produce a picture of a distant radio station by "Shearergraphing" the radio waves it emitted, he promptly obliged. Unfortunately, his dramatically clear picture was found to be identical with the frontispiece of that month's issue of *Wireless* magazine, and so the technique fell into disuse.

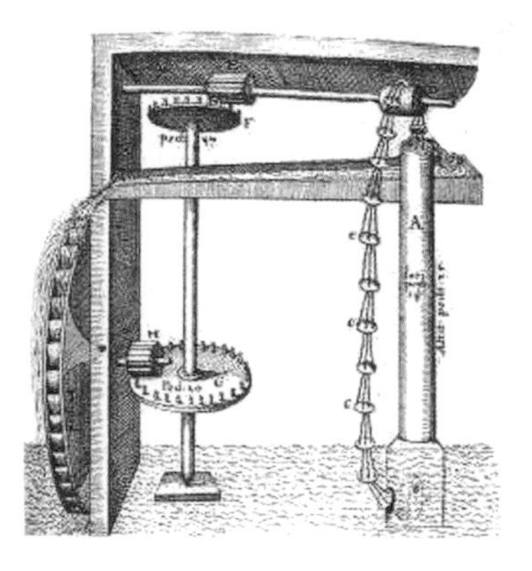

PERPETUAL MOTION

There are very good reasons why perpetual motion machines will not work. No machine can ever operate at 100% efficiency, so that further energy must always be supplied simply in order to keep the machine working. More fundamentally, it seems to be a fundamental law of the Universe that energy is conserved: in this context that means you can't get energy out of nowhere. With any machine, you've got to put more in than you get out.

The window on reality opened by quantum mechanics is also a window through which perpetual-motioneers can peer. One obvious attraction is the revelation that the hardest vacuum is not in fact empty. Rather, it is full of virtual energy – it can thought of as, if you like, a probability sea. The energy filling the emptiest vacuum is present in exactly equal amounts of what we can call positive and negative energy; the two exactly balance each other, so that the net energy value is zero. However, oppositely "charged" pairs of energy particles can – and do – appear spontaneously. In the ordinary way these mutually annihilate each other as near instantaneously as the human mind can conceive; however, there is probabilistically a minute chance with any particular pair-creation event that the annihilation will not happen. The energy is often called "zero-point energy" because it is present even at the Absolute Zero of temperature, where, according to all classical views of physics, motion ceases entirely and hence energy in any form becomes impossible. The future possibilities of vacuum mining are exciting to contemplate – at least in science fiction.

The most prominent predictor of the exploitable potential of zero-point energy is the US physicist Harold Puthoff (b1936), best known for his earlier work with colleague Russell Targ (b1934) advocating the commercial exploitability of paranormal powers. He expects that "vacuum engineering" will become the big white energy-resource hope of the 21st century, once we're out in space where there's all this exploitable vacuum just lying around for free. The 1979 Nobel Physics Prize laureate Steven Weinberg (b1933) has, unfortunately, poured cold water over the notion by pointing out that, in a volume of vacuum equal to that of the entire globe of the earth, there's about as much usable energy as there is in a gallon of petrol.

Allied to the notion of the perpetual motion machine is, as implied above, that of the antigravity device: without having to work against the immense pull of gravity, machines – and most especially aerial transport – would be immeasurably more efficient. In addition, the future of space travel could be revolutionized if spaceships didn't have to expend most of their payload in fuel, the vast majority of which is used up in working against gravity at the beginning and end of their journeys.

The most substantial body of work done towards devising an antigravity machine came about as a result of the obsession of the US businessman Roger Babson (1875–1967). During and after a successful career in business – he was the publisher of *Babson's Washington Service* – he turned to philanthropy. In our context his most notable achievement was the setting up in 1948 of the Gravity Research Foundation, a body specifically designed to discover means of reducing or entirely blocking the influence of gravity. The Foundation was based in New Boston, New Hampshire, a town selected by Babson as being sufficiently far from any major city that it was likely to survive a nuclear war.

The Babson Foundation held seminars that attracted even some fairly eminent scientists – such as Igor Sikorsky (1889–1972), designer of the first successful helicopter (1939) – but of considerably more importance were the annual essay contests it sponsored, drawing papers from all over the world on gravity-related subjects. As the focus of the Foundation's research slowly shifted from antigravity toward gravity in general, these essays began to assume considerable scientific merit; among the multiple winners of the contest has been Steven Hawking (b1942).

After Babson's death in 1967 the Foundation slowly wound down, and it's now essentially defunct – although the essay competition continues at least sporadically. At the time of writing the Foundation's website was operating in skeleton fashion while seeking a sponsor.

A superb thought-invention came from George Rideout of the Babson Foundation. *If only* there were a material which acted as a gravity shield – that is, if you stood on it you would no longer feel the pull of the Earth's gravity – the device could be built. Imagine a spinning bicycle wheel on a horizontal axis. Put a plate of the gravity-shield material under one side of the wheel (say, under the left-hand semicircle as the wheel faces you). Now consider two of the particles that make up the wheel, A and B; A is above the plate while B is diametrically opposite A, on the other side of the wheel. Give the wheel a clockwise twirl, and it will keep on spinning forever because gravity is pulling all the particles B downwards while no energy is required to raise all the particles A. The machine is, of course, using the Earth's gravity as "fuel" – in very much the same way that a water-wheel uses running water.

This is all very well, assuming the gravity-shield material existed; but it seems unlikely that it does. In fact, if it did, you could make a much simpler perpetual-motion machine by tying a plate of the stuff to your boots and jumping.

As one might expect, the Breakthrough Propulsion Physics Projects at NASA receives a large number of communications each year from amateur inventors convinced they have discovered the space drive that will solve all future propulsion problems and take humankind to our glorious destiny among the stars . . . None so far have been found to work, but, who knows, one day a breakthrough might indeed arrive unsolicited. In order to reduce the gargantuan amount of time that would be required to analyse each and every offering, NASA has compiled lists of *principles* that are known not to work. Most of the proposals that arrive involve one or more of these principles, and can therefore be discarded after no more than a cursory examination.

Among erroneous principles frequently employed by the submitted proposals, the three most common categories are gyroscopic antigravity, electrostatic antigravity, and oscillation thrust.

Gyroscopic Antigravity: The most famous gyroscopic antigravity device was one devised by the UK inventor Eric Laithwaite (1921–1997) – designer of the world's first high-speed train and of the Maglev, and Professor of Electrical Engineering at Imperial College, London – and was demonstrated by him to the Royal Institution in 1973. His device, which looked like a very large gyroscope on a pole, weighed about 25kg (50 pounds), and the first thing he did was show he could barely lift it. He next spun up the gyroscope using an electric drill and showed he could now lift the contraption above his head, one-handed. He then remarked, either in jest or seriously, that he was demonstrating a violation of Newton's Laws of Motion . . . and the Royal Institution was not amused, for the first and only time in its history refusing to publish an account of a lecture given to it.

Laithwaite was puzzled by the physics of a rotating gyroscope, which did indeed *appear* to be violating Newton's Laws, and spent many years probing the situation. Finally he was able to show mathematically that the Laws were not being violated; at the same time he still believed the behaviour of gyroscopes could be harnessed to produce a reactionless drive. Near the end of his life he applied for and was granted a US patent to exactly this effect. The fact that his reactionless drive has never been taken beyond the prototype stage has of course fuelled speculations that it might perhaps work, despite all the reasons to believe it wouldn't. Laithwaite was the first to announce that his drive requires fuel in sufficient quantities that it seems it would offer little advantage over other propulsion units.

There's some evidence from General Relativity that a gyroscope spinning at Relativistic speeds might indeed affect local gravity, but unfortunately the speed of the gyroscope's motion would also mean the device's mass was increasing towards infinity.

Electrostatic Antigravity: Devices employing what is called electrostatic antigravity typically involve passing a high voltage across an unusually shaped capacitor; the capacitor rises above the ground in a manner that resembles levitation. Various studies of the effect have concluded that the lift is given by what's called an ion wind: ions passing from one capacitor electrode to the other generate an air flow and, if the electrodes are suitably placed, this air flow can give the capacitor an upward thrust. There's some evidence that, even in space, there could be enough of an ion flow to give the capacitor some thrust. Unfortunately, there's the matter of the power supply, which weighs considerably more than the capacitor and has to be connected to it by a wire. So far no one's been able to envisage an ion wind powerful enough to move not just the capacitor but the power supply.

Oscillation Thrusters: The type example of the oscillation thruster is the infamous Dean Drive, which in 1960 grabbed the attention of John W. Campbell Jr (1910–1971), editor of the science-fiction magazine *Analog*; for a long time he ran "science fact" articles trying to persuade his readers that it was the breakthrough that would herald the age of interstellar travel. Devised by Norman L. Dean, a mortgage appraiser, this little gadget could lurch convincingly around Campbell's desktop and, if put on a bathroom scale, appeared to lose weight when turned on. It worked on the same principle as other oscillation thrusters: essentially, if you can rig a sequence of weights such that they shoot at high speed in one direction before returning slowly in the opposite direction to their starting point, you will indeed appear to generate thrust . . . assuming your gadget is on a surface like John W. Campbell's desktop. What's actually happening is that the jolts produced by the high-speed weights going in one direction are sufficient to overcome the friction between the gadget and the desktop, while the slower, gentler movements in the other direction are not; thus you get motion of the gadget as a whole in the "positive" direction. Unfortunately, there's no friction to speak of in space, and so all a gadget like the Dean Drive would do is oscillate around its starting position.

COLD FUSION

On March 23 1989 two scientists working at the University of Utah, Stanley Pons and Martin Fleischmann, announced they had discovered a technique which would make virtually unlimited energy available to humankind for the virtually unlimited future, and all at a staggeringly small cost. The process concerned was cold fusion.

The nuclear energy we use for electrical power today is the product of nuclear fission, the disintegration (decay) of large atoms into collections of smaller ones, with the release of energy. There are inherent dangers because, not only is the fuel radioactive, so are some of the by-products of the process. However, almost at the same time that physicists recognized the energy possibilities of nuclear fission, they saw that similar if not even greater rewards could be obtained by, rather than breaking up big atoms to make smaller ones, jamming smaller atoms together to make bigger ones. This process is known as nuclear fusion and is what keeps all of us alive: it is the process almost exclusively responsible for making the stars, including our Sun, shine.

At the simplest level, if you take two atoms of hydrogen, the lightest and simplest (and most abundant) of all the elements, and bang them together just right, you find you have one atom of helium, the second-lightest and second-simplest element . . . plus a certain amount of detritus that was present in the two hydrogen atoms but is not required for the making of the single helium atom. This detritus largely takes the form of energy. As with fission, so with fusion: you can use the energy for either bombs or usable power. The big differences between fission and fusion, in terms of applying them, are that fusion is "clean" – its by-products are harmless items like helium gas and water – and that its fuels are cheap and abundant. Control the fusion process and you've largely solved the world's energy problems.

Alas, nobody has yet been able to set up a controlled fusion reaction in any useful sense. Those that have been achieved have lasted for only fractions of a second and the power output has been many times less than required to make a flashlight flicker. And so there arose the dream of "cold" fusion – that is, nuclear fusion sustainable under conditions of heat and pressure not too dissimilar from the ones we're accustomed to. To show the viability of cold fusion as a technology we don't have to show that it works *well*, just that it works at all – that there is a genuine effect of more energy coming out of the endeavour than we had to put in. If this is the case, then straightforward human ingenuity can be brought to bear on the much simpler problem of making the process more efficient.

Stripped to the basics, the Pons/Fleischmann experiment depended on the known fact that the metal palladium has the property of "soaking up" the nuclei of deuterium ("heavy hydrogen"; where ordinary hydrogen has just a proton for nucleus, the nucleus of heavy hydrogen contains a proton and a neutron). To initiate fusion in a gaseous medium requires extremes of heat and pressure; with palladium, a solid, as a substrate, the constraints placed upon the deuterium nuclei are such that the concentrating effect is the same as if they were in a gas under intense pressure. Therefore it's worth at least checking to see if, by supersaturating palladium with deuterium, you can create conditions that might precipitate fusion reactions between the deuterium nuclei.* This Pons and Fleischmann accordingly did. They then measured extremely

* There are theoretical reasons why deuterium is a better material for effecting fusion than ordinary hydrogen.

accurately the temperature of the palladium and its surrounds to see if heat was being generated. Their results seemed to show that indeed heat was.

Physicists and chemists worldwide dropped everything to try to replicate the experiment. However, while the apparatus might be inexpensive and easily obtainable, the measurement of such minute energies was far trickier. Some who were unacquainted with the way that science is done – no experiment is considered valid until it has been duplicated and the results checked – were impatient with this confirmation process, and such hotheads included many financial speculators and a truly shameful number of politicians. The Utah State Legislature promptly hurled four and a half million dollars at Pons and Fleischmann. The US Office of Naval Research chucked in an initial $400,000. It looked as if at least tens of millions of dollars might be on the way from the US Government. When initial reports from other scientific researchers seemed favourable towards the Pons/Fleischmann results, further financing from industry seemed a foregone conclusion.

But trouble was brewing for the two chemists and their most ardent supporter, the University of Utah. While some early efforts elsewhere to reproduce the results had seemed to hint at confirmation, a disturbing number of others now did exactly the opposite, and soon the negative reports became the majority. Matters were not helped by some apparent belated fudging of figures on the part of a desperate Pons and Fleischmann in defence of their increasingly beleaguered position. There were also astonishingly crass attempts by the University of Utah (who promptly disclaimed responsibility when this matter became public) to use the weapon of threatened litigation to silence critics – more than anything else this destroyed Pons's and Fleischmann's credibility. (It's an obvious rule of thumb that only a scientific illiterate would attempt to use a lawsuit to influence a scientific debate.)

With the onslaught of adverse experimental evidence came demolitions of the theoretical underpinning of Pons's and Fleischmann's research. To take just a single example, it was shown that deuterium nuclei are actually further apart in saturated palladium than they are in heavy water; since heavy water does not spontaneously heat up because of fusion reactions occurring in it, why should palladium?

We shouldn't forget, however, that at least a few researchers believed they were able to duplicate Pons's and Fleischmann's results – and, in the years since the main furore died down, others have joined their ranks. The two Utah professors likely did not, as they thought, discover cold fusion; but they may have discovered *something* – and this something, whatever it might be, has not yet been fully investigated.

Similar doubts surround the claims made in 2002 by Rusi Taleyarkhan, then of the US Department of Energy's Oak Ridge National Laboratory, Tennessee, and later of Purdue University, Indiana, to have attained cold fusion. His team bombarded a beaker of chemically altered acetone with neutrons and then with sound waves to create bubbles; when the bubbles burst, the team reported in *Science*, fusion energy could be detected. Other teams, however, have had difficulty duplicating the results – including Taleyarkhan himself. Working at Purdue, he finally announced in 2004 that he had done so using the uranium salt uranyl nitrate. There have been many questions raised about this. Brian Naranjo of the University of California, Los Angeles, who in 2005 reported that his team there had attained cold fusion by heating a lithium crystal soaked in deuterium gas, analysed Taleyarkhan's results and concluded that the Purdue scientist had detected not cold fusion energy but leakage from some other radiation source present in the laboratory. If so, it seems a rather elementary mistake for Taleyarkhan to have made. More gravely, some of Taleyarkhan's Purdue colleagues began raising complaints about various aspects of the experiment, saying that he claimed to have achieved positive results in experimental runs for which he declined to produce the raw data, that he opposed their publication of their own, negative results, and so forth. At the time of writing his conduct is being reviewed by Purdue University.

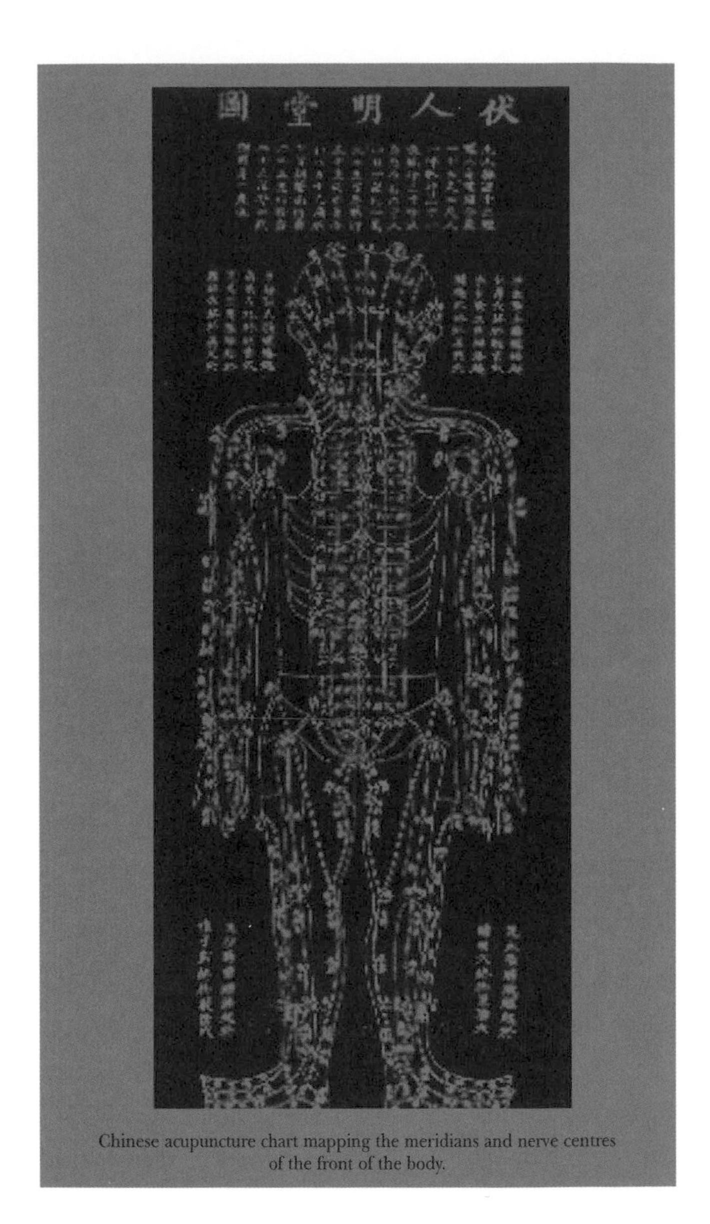

Chinese acupuncture chart mapping the meridians and nerve centres of the front of the body.

CHAPTER 6

Us... Or Something Like Us

According to Aristotle (384 322BC), there is in living creatures a fundamental vital principle, a "life force", which distinguishes them from nonliving material. The gods breathed this vital principle into living things, and thereby gave them their life – a workable explanation for Spontaneous Generation (see page 125). The early alchemists seem to have thought this principle so real that, not only did they consider all entities to be made up of differing proportions of dead matter and life force (spirit), they attempted to employ it much like any other chemical in their experiments – they virtually tried to bottle it. This idea, vitalism, was still a respectable theory in the hazy days when qualitative alchemy was being transformed into quantitative chemistry: Georg Ernst Stahl (1660–1734) was one of its supporters.

The theory started to fall to pieces in 1828 when Friedrich Wöhler (1800–1882) was able to synthesize urea, a major waste product of the body and thus indubitably organic, from inorganic materials. In the 1840s Emil Du Bois-Reymond (1818–1896) proved experimentally that the impulses which travel along the nerves fit into the mechanistic scheme of things, being akin to electric currents; and in 1894 Max Rubner (1854–1932) found that the amount of energy which the body extracts from food can be predicted by the laws of thermodynamics. The 1896 discovery by Eduard Buchner (1860–1917) that fermentation does not require the presence of living cells was merely the final nail in the coffin.

Yet, vampire-like, the theory refuses to stay in that coffin. In the mid-19th century Karl von Reichenbach (1788–1869) came close to it with his "odic force". Odic force, od or odyle is rather like an electromagnetic field permeating everything, and was invoked by him to explain most paranormal phenomena. It is of interest here in that he thought the human body is rich in luminous od: care had to be taken in the selection of clothes lest one don something made from an od-inhibiting fabric. Spirit mediums could see the aura of od surrounding a person's body. This is not far distant from the aurae which practitioners of Kirlian photography claimed to be able to capture on film.

NOUS

In the most simplistic sense, nous was a vital principle somewhat akin to the life force of vitalism: the nearest English words to describe the Greek concept are "mind", "intellect". But it's important to realize that the range of the word "nous" is vastly greater than that of our word "intellect", its precise meaning varying with both context and user.

On the one hand, it can simply mean intelligence (as in "using your nous"); more precisely, it refers to the combination of insight and intuition which permits the apprehension of the fundamental principles of the cosmos. Thus the concept is closer to the oriental idea of "seeing" than to the occidental notion of intelligence founded upon Euclidean logic.

At the same time, nous could also be the creative, motive intelligence behind the cosmos, almost indistinguishable from the Christian concept of the will of God. In the cosmology of Anaxagoras (*c*500–*c*428BC), the Universe began as a homogeneous sea of identical basic particles. Nous gave this sea a quick stir, in the knowledge that over time the particles would so combine and arrange themselves that everything would be the way it is today.

THE HUMOURS

The mediaeval system of the four bodily humours, which gave rise directly to the medical school of enantiopathy and less directly to that of allopathy (see page 279), was born from a desire to see in bodily conditions the attributes of the four Aristotelian elements (see page 240). Thus earth was reflected in the body as black bile or melancholy; air as yellow bile or choler; fire as blood; and water as phlegm. We still use the words "melancholic", "choleric", "sanguine" and "phlegmatic", not to mention "bilious". If the individual were hot and dry, then he was suffering from an excess of blood (equivalent to fire, which is hot and dry) and therefore was leeched – and so forth. Mortality rates were high.

IATROCHEMISTRY

While the concern of the mainstream of alchemy was the attempt to transmute base metals into gold and silver, in the 16th century there emerged a school which brought the techniques and philosophical theses of alchemy to bear upon the preparation of medicines. To the traditionalist the symbol of alchemical purity was gold; to the iatrochemist it was the ultimately pure medicine. Thus iatrochemistry was really a forerunner of pharmacology.

The two main figures involved were Paracelsus and Jan Baptista van Helmont. Van Helmont (1577–1644) made important contributions in various fields of science. He was the first to realize there are gases other than air, and actually coined the term "gas"; yet he remained a fervent believer in alchemy and in the power of the Philosophers' Stone. One of his spurious hypotheses – that matter is made up almost entirely, if not entirely, of water – produced an important and worthwhile experiment. He weighed a quantity of soil and in it planted a tree-shoot. Five years later he found that the weight of the tree had increased by over a thousandfold, while that of the soil had decreased only slightly. Clearly the tree's increased weight had been given to it almost exclusively by his watering of it, and hardly at all, as others believed, by the soil. His conclusion, that the water was somehow transmuted into tree-stuff, was unfortunate (he didn't think that air might play a part, for instance), but the experiment had at least scotched one current fallacy.

But the central figure in iatrochemistry was undoubtedly Paracelsus. His writings were extremely – and deliberately – obscure (and often obscene), so it is frequently hard to determine what he was actually saying. Thus one is in the uneasy position of not knowing which of his ideas were erroneous, which intentionally misleading, and which before their time. For example, it was long, long after his death that it was found that his instructions for an "extract of vitriol", which he had certainly used to sedate chickens and almost certainly in his medical efforts, were in fact those for making anaesthetic ether. Similarly, his pioneering use of metals in therapy was not to be accepted for centuries (although in some cases, in a pre-echo of homeopathy, his metallic distillate was so fine that there was no longer any metal left in the potion).

He did important early work on several diseases, notably syphilis, although he thought syphilis was merely a sort of symptom of other diseases, rather than a disease in its own right. Equally important, in an age when the cure could be far more dangerous than the disease, was his caution concerning the *amount* of treatment to use: he recognized that our bodies have their own self-healing abilities, although much of these he attributed to tissues containing an active healing principle, *mumia*.

Paracelsus replaced the four Aristotelian element by three "hypostatic principles": mercury, sulphur and salt (mercury and sulphur were already alchemical favourites). While by "salt" was meant *very* roughly what we would think, the same is not true of his mercury and sulphur. These three were principles common to all living things. When distilling an organic substance, one would find the first to evaporate would be the thin, volatile constituent, the "mercury". It acted in favour of life, of youth, and of form-change. Next came the sticky stuff, the "sulphur", which acted in favour of, for example, growth and increase. Finally one would be left with a dry mass at the bottom of the vessel; this was the "salt".

Clearly the concept of these three principles – even when the number was later increased – was only a slight advance upon that of the four elements, especially when elaborated with occult mystery. And so Paracelsus not only initiated the modern science of chemistry, he also launched it on what was to be both a long and a wasteful digression, not to be ended until the "fall" of phlogiston.

SEX

It is illegal to seduce the wet nurse of an heir to the British throne. You might taint the royal blood with your plebian characteristics, which would be transmitted in the milk to the infant. In similar vein, the Ayatollah Khomeini (*c*1900–1989) ordained: "If one commits an act of sodomy with a cow, a ewe or a camel, their urine and their excrements become impure, and even their milk may no longer be consumed."

From ancient times until at least the last century, the theory has frequently surfaced that sperm from the right testicle produces boys, that from the left, girls (some authorities had it the other way around). Aristotle proposed a simple experiment: remove one testicle and see if further children born to this father thereafter are all boys or all girls.

Talking in 1975 of a National Science Foundation grant for researches into the nature of love, Senator William Proxmire (1915–2005), famous as the antiscientific scourge of NASA, came out with one of the more inaccurate generalizations of recent decades: "Two hundred million Americans want to leave some things a mystery, and right at the top of those things that we don't want to know is why a man falls in love with a woman and *vice versa*."

Sex is a wondrous playground for unorthodox theorists. One gasps in admiration at the sheer imaginative scope sometimes involved. Sir Richard Burton (1821–1890) discovered that homosexuality is concentrated in the Sotadic Zone, a great belt girdling the world. The title of *Did Man and Woman Descend From Different Animals?* (1927) by William Smyth is fairly self-explanatory, and the theory was debated as recently as 1980 in the pages of the *British Medical Journal*. Hippocrates (*c*450–*c*370) realized the Scythians had died out because they were all impotent from spending long hours in the saddle. Pope John Paul II (1920–2005) in 1980 pronounced that, not only is looking lustfully at someone else's spouse a form of adultery, looking lustfully at your *own* spouse is adultery, too. Controversy in the Islamic world was caused in early 2006 by a fatwa issued by the Egyptian cleric Rashad Hassan Khalil to the effect that married couples should keep some clothes on during sex: if they made love naked their marriage was invalid. Other Islamic scholars, not to mention the vast majority of the Islamic public, disagreed forcefully. Barbara Cartland (1901–2000) proposed that young brides might wish to wear a pair of white gloves on the first night of marriage, just in case they had to touch the beastly thing. Arabella Kenealy (1859–1932) wrote *The Human Gyroscope: A Consideration of the Gyroscopic Rotation of Earth as Mechanism of the Evolution of Terrestrial Living Forms: Explaining the Phenomenon of Sex: Its Origin and Development and Its Significance in the Evolutionary Process* (1934), another work whose title really tells you as much as you need to know. Edward Clarke (1820–1877) in 1873 warned that education might cause a woman's uterus to shrivel up. The list of spurious sex-related hypotheses is endless.

A frequent theme is that sex is permissible so long as you don't enjoy it – like the Victorian ladies who were advised to lie back and think of England. But then the Victorians were concerned even about the dangers of kissing. In *What Everyone*

Knew About Sex (1972), William Dwyer quotes the US sexologist Orson Fowler (1809–1887): "When you kiss and allow yourself to be kissed *with an appetite*, to fondle and be fondled, hug and be hugged, you are thereby perpetrating mental sexual intercourse." Similar Fowlerian strictures applied to the sex act itself, even within marriage: the child born of a wild night of love (especially if the spouses were tipsy) would probably be depraved, mentally subnormal, or, worst of all, "like Satan". Curiously, it was according to Fowler at the same time dangerous *not* to enjoy the act of love, since children born of dispassionate bouts might well be sickly and weak in the head. Presumably one had to strike a happy balance between drudgery and ecstasy.

To these 19th-century physicians, then, even sex within marriage was fraught with dangers. Some modern sexologists suggest that on the first night of the honeymoon sex might be inadvisable, since both partners may be tired and/or drunk, but some of their predecessors suggested waiting for as much as a year after the marriage day. For the first couple of decades of their lives young ladies had been trained to be chaste: the shock of being all of a sudden "allowed to" might well lead to lunacy.

Of course, sex lay at the heart of most if not all of the old religions. Our ancestors were fundamentally concerned about the annual agricultural cycles of fertility, and so sex – a triumphant declaration of fertility – quite naturally formed a focus for their religious thought. In some rural areas, such customs still survive as that of settling the newly planted crops by, well, banging them down into place. Strange, then, that Christianity in particular has had such a long history of disapproval of sex – certainly of fornication and often, as noted, even of the enjoyment of marital sex. This dislike, together with a strong consciousness of the fertility aspects of sex (remember, Onan's sin was not masturbation but the wasting of his seed), has had dire consequences over the centuries. Sex acts indulged in for pleasure, without any possibility of fertilization occurring, have generally been regarded as crimes –

homosexuality, masturbation, oral sex, sodomy, even intercourse with a sterile spouse. But Herbert W. Armstrong (1892–1986), for decades Editor-in-Chief of *The Plain Truth*, in its April 1981 issue reassured Christians everywhere that the ghastly sin of actually enjoying sex would soon cease to be a problem: "And those mortals who do reach their incredible HUMAN POTENTIAL . . . will no longer have sex organs – no longer male or female but . . . as the angels in heaven, no sex differences."

The idea of the homunculus sprang from a theory called preformationism (or ovism, or animalculism), which had it that all the major structures of the eventual adult are already present in one of its parent's sex cells – either its mother's ovum or its father's spermatozoon. The origins of that theory can be traced to the discovery by Regnier de Graaf (1641–1673) of what he believed to be the mammalian egg (in fact it was a misidentification: another couple of centuries would pass before instrumentation became adequate for mammalian ova to be observed). A short while later Antony van Leeuwenhoek (1632–1723) discovered the male sex cell, or spermatozoon, which wriggled like an independently living thing, or animalcule.

It was the Dutch microscopist Nicolaas Hartsoeker (1656–1725) who suggested that each spermatozoon contained a tiny adult. The process of foetal growth from the homunculus was regarded as being similar to the growth of a baby to adulthood: while there were changes of form and size, there were no overall morphological alterations – unlike the case with, for example, caterpillars and moths. Thus in each sperm cell or ovum there was curled up a minute human embryo, the homunculus; if we are to believe some of the artists' representations of the theory, the homunculus could alternatively ride astride the spermatozoon much as a witch might ride a broomstick.

The early microscopists, peering hopefully through their

imperfect lenses, straining at the limits of visibility, were actually able to *see* homunculi – in much the same way and for almost exactly the same reasons as Schiaparelli was able to see "channels" on Mars through his telescope (see page 72). One naturalist, Jean Fabien Gautier d'Agoty (1747–1781), was able to display, suspended in a glass of water, a specimen of a homunculus large enough to be visible to the naked eye. But not all scientists were so self-deluding. They noted, for example, the contents of a hen's egg cannot be said to look anything like a chicken. Rather than abandon preformationism, they pointed out, quite correctly, that just because the yolk did not *look* like a chicken did not prove there was not a miniature chick embryo present there, but too tiny and transparent to be visible.

A major flaw in the theory was that, if indeed the homunculus contained in rudimentary form all the attributes of the eventual adult, then those must include the adult's sex cells, which would themselves necessarily have to contain minute homunculi. And those homunculi, too, would have to contain yet tinier homunculi . . . and so the progressively smaller populations of homunculi would have to continue *ad infinitum*, or at least for the number of generations that God was going to permit mankind to inhabit the Earth. In fact, to the naturalists of the 17th and 18th centuries this did not seem as ridiculous as it does to us today, because they had no knowledge of cells and hence no conception that there could be a lower limit to the size that a complex structure like a homunculus could have.

Onan's crime was seed-wasting, and thus was a moral offence not because of the self-pleasuring but because of the betrayal of the Jews, whose God-given duty was to have as many children as possible. For centuries the Christian Church has frowned upon sexual activities from which pregnancy could not result: even today, the Roman Catholic hierarchy maintains a stubborn opposition to contraception, on the basis that the act of love should be merely an act of procreation. In comparatively recent times, Church, state and medical science combined to condemn all forms of seed-wasting as dangerous to body and/or soul. Recently the spread of enlightenment has been reflected at most levels in Western society, although some populist bigots still lead verbal and even physical attacks upon, especially, homosexuals, singled out for vituperation by right-wing US televangelists at any excuse, or even without one. In the wake of the September 11 2001 Al-Qaida terrorist attacks on the US, the Reverend Jerry Falwell (b1933) pronounced:

> The abortionists have got to bear some burden for this because God will not be mocked. And when we destroy 40 million little innocent babies, we make God mad. I really believe that the pagans, and the abortionists, and the feminists, and the gays and the lesbians who are actively trying to make that an alternative lifestyle, the ACLU, People for the American Way – all of them who have tried to secularize America – I point the finger in their face and say, "You helped this happen."

Exceptions to the general Churchian rule are the inefficient contraceptive techniques of *coitus interruptus* and *coitus reservatus* – in the latter, the male avoids orgasm altogether. This was the method of contraception used in the Oneida Community (where polygamy and polyandry were also practised). At the end of the last century Alice Bunker Stockham (1833–1912), in *Karezza*, recommended the practice, at the same time suggesting that intercourse should take place at most every few weeks, preferably only every few months.

But masturbation was a real sin. One popular Victorian theory was that a man could discharge only so much semen during his lifetime. This theory was used by doctors and laymen alike to frighten boys away from self-abuse. Worse things than running out of semen could happen, and in the shorter term. Fowler tells us: "Victims of self-abuse have pallid, bloodless countenances, hollow, sunken and half-ghastly eyes, with a red rim around the eyelids, and black-and-blue semicircles around the eyes. Red pimples on the face, with a black spot in their middle, are a sure sign of self-pollution in males, and irregularities in females . . ." A hypothesis based, no doubt, on the fact that most adolescents have acne and most of them masturbate: perhaps acne increases lust? "Masturbation poisons your body," adds Fowler, "breaks down your nerves, paralyzes your whole system. When practised before puberty it dwarfs and enfeebles the sexual organs. It also corrupts your morals and endangers your very soul's salvation! You may almost as well die outright as to thus pollute yourself."

You might have no choice here, for even "Doctor Napheys, although he believes that occasional indulgence may not leave after it those permanent effects . . . acknowledges that it often does lead to insanity or serious illness, and even early death." Since George Napheys (1842–1876) was a leading gynaecologist of the time, his opinion carried some weight. Napheys's reputation has suffered rather considerably since those days, and it's not hard to understand why. To choose just one example from his various writings: ". . . only in very rare instances do women feel one tithe of the sexual feeling which is familiar to most men. Many of them are entirely frigid, and not even in marriage do they ever perceive any real desire."

Detecting that your child had discovered the forbidden joys of masturbation was, according to Fowler, made yet easier by other signs. Offenders often sat with their guilty hand in their laps, thrust their pelvis forwards when laughing, and walked oddly. Female masturbators – who ran the risk of insanity, consumption, flat-chestedness and other forms of general doom – could be spotted because they were constantly crossing their legs.

According to the doctrine of focal sepsis (see page 290), surgery could be effective in curtailing the habit: excision of the ovaries was found to be effective, as was enlargement of the anus. For boys, castration was often proposed but rarely carried out.

Sodomy, either with another man or with a woman, is still a crime in many parts of the world. It is, of course, illegal in some states of the USA, where, paradoxically, to judge from the outpourings of the sexual crusaders, it is more popular than anywhere else in the Western world. Whether or not sodomy is physically injurious is still a matter of debate. Such doubts do not surround some of the other practices outlawed in parts of the USA. Several of the US states still have laws against unmarried sex, although the penalties vary widely. Strictest is Michigan, where the penalty could be up to a $5000 fine and five years in jail. In Arizona the threatened penalty is a three-year jail term. Branchville, South Carolina, threatens a $500 fine and six months in jail. In Texas it's just a fine of $500, while in Rhode Island the fine is a mere $10 apiece – great value.

There are also draconian penalties in several states for committing oral sex, even between married couples. Florida leads the charge, with a maximum sentence of twenty years, while South Dakota isn't far behind, with ten years. In Rhode Island it's seven to ten years; in New Mexico two to ten years plus a $5000 fine; and in Utah it's a six-month sentence plus a $299 fine.

Of course, such laws are in general simply ignored (except, for example, in sexual assault cases) until someone gets round to repealing them, but in 1988, in Clayton County, Georgia, Superior Court Judge William Ison instructed jurors hearing the case of Atlanta carpenter James David Moseley that consensual oral sex between Moseley and his wife, in the privacy of their own home, was a felony, and sentenced Moseley to five years in prison. The state's Parole and Pardons Board, reviewing the case, insisted Moseley should serve at least two years of his sentence – this at a time when Georgia's prisons were so overcrowded the state had to release 3000 convicted felons. Clive Stafford Smith, the international defence lawyer who more usually focuses on capital cases, took up Moseley's cause. When Moseley arrived in jail "the other prisoners, some of them murderers and such, burst out laughing," commented Stafford Smith.

Attitudes to homosexuality have varied over the centuries: the Greeks respected it; the Arabs gave us "a woman for duty, a boy for pleasure, and a melon for ecstasy"; yet in some parts of the world today it is heavily frowned upon, and in instances harshly punished, up to and including the death penalty. Shamefully, scientists have been little better in their prejudices. The first two editions, in 1952 and 1968, of the American Psychiatric Association's *Diagnostic and Statistical Manual of Mental Disorders* listed among the mental ailments officially recognized by that august organization "sexual orientation disturbance" – homosexuality, in other words. It was only after a militant political campaign that the APA agreed to reassess homosexuality's status; even then, although the APA's Board of Trustees voted unanimously in favour of declassifying homosexuality, when the matter was put to the vote of all the APA's members in 1974, fewer than 60% of the psychiatrists eligible to vote were in favour, the remainder presumably still believing that homosexuals were mentally ill.

Male homosexuality is the customary target of moral, legal and pseudoscientific prejudices. Lesbianism is a different matter. Queen Victoria (1819–1901) was expressing a fairly popular view when she decreed that sex between women was impossible.

From the heady world of sexual politics let us return to the more mundane problem of nocturnal emissions. In Victorian times, these were known to sap the strength and damage the health; moreover, what was nocturnal emission if not a subconscious form of the dread masturbation? Some sufferers – such as Frank Harris (1856–1931) – got around the difficulty using a tightly knotted piece of string, while others applied strips of sticking plaster to the offending member. Some doctors felt the disgusting event was a result of sleeping on one's back, and so recommended going to sleep with a towel around the waist, knotted at the back of the spine, to preclude the possibility of unconsciously turning over to lie supine. Most effective of all was the "spermatorrheal ring", which had sharp points on its inside. It was placed around the base of the penis last thing at night. The onset of an erection would awaken the wearer in agony. What the luckless wearer was supposed to do next is unclear.

One of the more curious sex-related conditions is that of *koro*, found in various parts of Asia. The *koro* sufferer becomes convinced that his or her (usually his) sexual organs are shrinking – the term *koro* is thought to come from various Malaysian and Indonesian words meaning "tortoise", since a slang term in that part of the world for the tip of the penis is "tortoise's head". If there are individual cases of *koro* they have been little documented: most often *koro* arrives in outbreaks, with hundreds if not thousands of people rushing to doctors and hospitals to have their penises (or breasts or vulvas) cured of the shrinkage that the patient, and often his family, can observe. In more extreme instances the sufferer is obsessed with the notion that his penis is retracting into his body and that, should it do so entirely, death will be the result. Needless to say, the investigating physicians can find no physiological change in the suspect penis at all, the only treatments necessary being in consequence of some of the more bizarre methods deployed by the sufferers in their attempts to stop their genitals retracting entirely – up to and including safety pins.

No one is entirely sure what causes *koro* outbreaks, but that they happen every few years in one part of Asia or another is undeniable; furthermore, they're by no means confined to the uneducated, although the uneducated usually form the bulk of the patients. The ailment seems to be anxiety-related: the shrivelling of the sex organs is of course one of the basic fears, and anxiety itself often results in a (temporary) shrinkage of the penis, thereby possibly offering some visible "proof" that the process feared by the sufferer is indeed underway. Once a few individuals in an area become convinced they're suffering from *koro*, then the panic spreads so that lots of other people believe they too are suffering the symptoms, in much the same way that, if one person tells the world s/he's seen a UFO, suddenly others start thinking they see them too – indeed, *koro* outbreaks bear all the hallmarks of UFO flaps.

In the US during 1943 there was particularly high absenteeism among women working on the filling of fire extinguishers. When sociologists tried to find out why, they discovered the rumour was rife that carbon tetrachloride could cause pregnancy. This is only one of the many useful excuses to hand should a woman find herself pregnant while her husband is off at the wars. Most extreme is to claim that she was crossing a field when a snake leapt up from beneath and penetrated her, in which case, alas, the child will be the epitome of evil. This was the rôle of the serpent in Eden, according to versions of the Old Testament earlier than about 500BC: after that the creature only tempted Eve.

There is an ancient theory that the mental experiences of the pregnant mother may affect her eventual child, as might parental emotions during conception. Thus we find in *Genesis* the ploy of Jacob, who has promised to look after Laban's flock on the proviso that, on Laban's return, Jacob may keep all the animals which are spotted, speckled, striped, etc.

> Jacob gathered branches in sap, from poplar, almond and plane trees, and peeled them in white strips, laying bare the white on the branches. He had put the branches he had peeled in front of the animals, in the troughs in the channels where the animals came to drink; and the animals mated when they came to drink. They mated therefore in front of the branches and so produced striped, spotted and speckled young. As for the sheep . . . he turned the animals towards whatever was striped or black in Laban's flock. Thus he built up droves of his own which he did not put with Laban's flock. Moreover, whenever the sturdy animals mated, Jacob put the branches where the animals could see them . . . But when the animals were feeble, he did not put them there; thus Laban got the feeble, and Jacob the sturdy, and he grew extremely rich, and became the owner of large flocks, with men and women slaves, camels and donkeys.

Manufacturers of candy-stripe sheets cannot know the distress they may have caused.

The theory was promulgated, authoritatively, as late as the 19th century. Orson Fowler told of a mother-to-be who during pregnancy yearned for grapes and was attacked by a turkey:

the child was born with a "large cluster of globular tumours growing from the tongue and exactly resembling our common grapes. And on the child's chest there grew a red excrescence exactly resembling a turkey's wattles." There could be worse. Fowler reports also a child born with an extra thumb, so that it and the real thumb together resembled a lobster's claw. This was because the mother-to-be had bought a lobster, which had afterwards been stolen. One woman foolishly went fishing while pregnant: her child was half-human, half-fish.

As an aside, the expression "lick into shape" comes from a long-held popular theory that baby animals are born formless, and literally have to be licked into shape by their mothers. While according to Lord James Monboddo (1714–1799), who maintained also that orangutans are members of the human race (and so provided the basis for "Sir Oran Haut-ton" in Peacock's *Melincourt*, 1817), all human infants are born with tails, which are deftly removed by the midwives.

A considerable body of popular fallacy surrounds menstruation. In some parts of Australia the Aborigines maintain that, if a man goes near a menstruating woman, he will lose his strength and grow prematurely old. Similar myths are alive in Western society. Eating ice cream during one's period is risky – but not as bad as washing your hair. Bathing during a period can cause tuberculosis. Flowers wilt and crops are blighted as menstruating women walk by. Even so, menstrual blood has been used medically. For long it was thought that fresh human blood was efficacious against leprosy, menstrual blood best of all; until quite recently it was recommended that the patient mix some in with the bathwater. Menstrual blood was also useful in the treatment of warts.

Bizarre theories concerning sex and sexuality are, as you will by now have gathered, legion, and new ones emerge with disturbing frequency. This has been the briefest of surveys.

Killing and Curing

While a large part of therapy has always depended on straightforward experiment, most of the great figures of medicine have realized that this is not enough: until we know the *causes* of disease, then therapy will always be a hit-or-miss affair. Some suggested causes are less mundane than others. For example, Athanasius Kircher (1602–1680), a pioneer microscopist, around the middle of the 17th century put forward the idea that the root cause of Plague was rotting mermaids.

Another theory, popular for millennia, is that bad smells cause disease. During the Plague years sensible people carried fragrant flowers or herbs to drown the malodour and so escape infection – the expression "a pocket full of posies" in the song (which is supposedly all about the Plague) refers to this habit. Again, the name "malaria" is literally translated as "bad airs"; for a long time it was thought that foetid air was the disease's cause – the sort of air, in fact, that you might breathe when walking around at night near a swamp in a steamy climate. (It used to be a popular theory that night air *per se* was bad for you.) It was not until 1898 that Sir Ronald Ross (1857–1932) of the British School of Tropical Medicine established that malaria is caused by female *Anopheles* mosquitoes inadvertently injecting the microorganism *Plasmodium* into our bloodstream as they feed. And, of course, *Anopheles* mosquitoes like living in swamps, and feed chiefly at night-time.

Christian Hahnemann (1755–1843), the "father of homoeopathy", coined the term "*Allopathie*" to describe those therapeutic techniques which, unlike homoeopathy, attempted to cure disease by inducing in the patient symptoms different from those which he/she displays. In fact, the idea of allopathy goes back a long way before Hahnemann, to the Middle Ages, when medical science believed that our physiological welfare was governed by the four humours (see page 265). In mediaeval times, since theoretical knowledge of disease was rudimentary at best, treatment could be based only upon the outward signs of ailment, the symptoms. There thus grew the belief that to subdue the symptoms was to subdue the disease. This scheme was called enantiopathy: cures were attempted by applying the opposite of the symptoms. This not unreasonable idea was fatal when applied in a medical system based upon the humours. Consider a person suffering from influenza. He is hot and thirsty (i.e., dry). The obvious answer is to treat him with a mixture of cold and wet. So flu patients sat in bathtubs of cold water in draughty rooms, waiting for their cure.

Much of our modern medicine is of an allopathic nature – but not all: we tend to keep people suffering from fever warm, not cold. And now that we have a greater understanding of the processes responsible for illness we can recognize which symptoms should be ameliorated and which should be left alone. Thus, while the taunt of the homoeopathist is that we rely too heavily upon allopathic treatments, the truth is that we rely upon them only when they actually work.

Another ancient technique was leeching. This involved the strategic placing of these bloodsuckers on the patient, so that the creatures would suck some of his blood; this they did with such a zeal that a medicinal leech could survive for up to a year on one meal. The practice gave rise to doctors' being dubbed "leeches". In medicine based on the humours, the practice must have seemed reasonable enough – what better way of correcting an overabundance of blood? But it is strange that the height of the technique's popularity was as !ate as the 19th century, when leeching was used for madness, whooping cough, gout, headaches . . . In fact, so popular was the practice that in some areas the leech species concerned, *Hirudo medicinalis*, became extinct through "over-cropping".*

The use of gold in medicine dates back even further, to the ancient world; it was fundamental to the alchemy-derived iatrochemistry (see page 265), of course; and there were sporadic reports of gold-based remedies even into the 20th century. The sole scientific basis for such medical deployment of a substance whose value rests primarily in the fact that it is

* Leeching has made something of a comeback in recent years, although with results that have yet properly to be evaluated.

Illustration from Paracelsus's *Surgery* (1549) showing a leg amputation. Paracelsus was a pioneer in recommending anaesthetics for such operations.

chemically fairly nonreactive seems to have been that it was such an excellent metal that it surely must have therapeutic virtues in addition to all its others. Only pure gold should be used; for ingestion, it worked better if prepared by heating it over a slow fire. Most physicians administered it by adding small quantities of gold to a potion, "potable gold", concocted from various herbs which in themselves had some curative properties; in other words, while the gold in fact played no part – favourable or adverse – in the subsequent events, the potion as a whole might benefit the patient, thereby perpetuating the myth.

External applications were believed to be beneficial, too. Gold would draw the blood to itself, and thus might be expected to prevent or heal, for example, the ugly pockmarks of smallpox. As late as the 18th century, wealthy people frightened of smallpox would pay handsomely to have their faces covered in thin gold leaf in order to prevent the formation of these blemishes. Kelemen Mikes, secretary to Prince Francis Rákóczi (1676–1735), wrote in 1718 about the unfortunate experience of one Countess Bercsenyi. Apparently the physicians had no difficulty in gilding the good lady's face; the problems started later, when they began to try to get the stuff off. Finally they had to resort to peeling it away piece by minutely small piece, using needles. Even then they couldn't remove the gold entirely from her nose, which for the rest of her life remained black.

Quacks have played a relatively minor rôle in the formulation of spurious medical theories: their contribution has been primarily to the corpus of useless cures.*

The emergence of medicine from a collection of often lethal superstitions and old wives' tales to become a science can perhaps be best exemplified by Ignaz Semmelweis (1818–1865), who demonstrated the power of the Scientific Method, even when practised in ignorance. In the 1840s, before medicine understood anything of the role of microbes

* They are thus considered in a separate, forthcoming volume on the corruption of science.

in disease, in one of the maternity wards of the Vienna General Hospital the maternal mortality rate from puerperal fever was horrifyingly high, at about 11%; further, this mortality rate was about six times as high as in the other maternity ward. Nobody could think why: the various guesses offered by the medical staff could be summed up as, basically, "bad vibes". The situation was complicated by the fact that there was a social difference between the two wards: the women in the affected ward came from the lower classes – although this might lead one to expect a lower mortality rate there, as strong working women should surely be more likely to survive than the delicate flowers in the other ward.

One of the physicians in attendance, Semmelweis, took a scientific approach – this despite opposition from his immediate superior, who believed the mortality rates were inevitable and so the investigation was a waste of time. Realizing that in the second ward he had what was in effect a control group (even though the mortality rate there, 2%, was miserably high by modern standards), Semmelweis analysed the differences between the two wards and their patients, and on the basis of that analysis advanced four hypotheses, and proceeded to test them in turn: women were encouraged to change their delivery position, the crowding was eased, and so on. None of these measures made any difference. Then, by chance, a friend of his cut himself while performing an autopsy, and thereafter sickened and died. Semmelweis put this datum together with the fact that the hospital's trainee doctors were the ones most likely to have been handling corpses – dissections were an important part of their training – and for his fifth hypothesis posited that there was some "cadaveric material" that could cause fatal illness. He tested this by insisting that everyone wash their hands before attending the mothers.

The mortality rate plummeted and Semmelweis was vindicated. That didn't save his job: he was soon fired by the hospital, in 1849, because of his radical politics. What's especially interesting, though, is that the explanation he derived from his hypothesis-followed-by-experimental-testing approach was both right and wrong: there *was* toxic "cadaveric material" being transported from the corpses to the unfortunate mothers, but not any kind of material that Semmelweis could envisage.

One might have thought his idea of hygiene in hospitals would have spread rapidly once his results became more widely known, but in fact the medical profession of the time pointedly ignored him: he was, after all, just a humble physician, a man without any particular standing in any of the important medical societies. Back in his native Hungary, he practised hygiene in his own obstetrics with enormous success, collected plentiful evidence of its efficacy, and even, in 1861, published the book *Die Ätiologie, der Begriff und die Prophylaxis des Kindbettfiebers*, copies of which he sent to all the medical societies; it too was ignored. Meanwhile, hundreds of thousands of mothers and children died all over Europe; in autumn 1860, in the very ward at the Vienna General Hospital where Semmelweis had reduced the mortality rate to 1%, that rate was almost 35%.* The horror of what was going on despite the knowledge he had unearthed understandably drove Semmelweis into chronic depression; his friends, assuming his obsession with clinical hygiene was of a piece with his other eccentricities, had him forcibly committed to a mental asylum, where he died in squalor.

At about the same time that Semmelweis died, Joseph Lister (1827–1912) began experimenting with surgical antisepsis and very soon discovered that it worked. Lister was better professionally connected than Semmelweis, but not hugely so: he was a Scot, working in Scottish hospitals, so for well over a decade the medical bigwigs in London continued to ignore him, too. All told, a quarter of a century was wasted – along with an untold number of lives – for no better reason than professional arrogance.

Despite this hiatus, medical theories began to resemble

* Another voice in the wilderness recommending surgical hygiene was the US physician Oliver Wendell Holmes (1809–1894), best known to us now as an essayist; he too was ignored.

reality, a little. The work of Louis Pasteur (1822–1895) helped establish the "germ theory" in the latter part of the century. Not everyone agreed with him, though. Rudolf Virchow (1821–1902), who in 1885 would make the important breakthrough of realizing that bodily cells reproduce – the death-knell for Spontaneous Generation (see page 125) – suggested that disease was basically *chemical* in nature. He had found that diseased cells develop from healthy ones, but refused to believe this might be due to the influence of a germ. (Later in life he refused to believe in Darwinian evolution – see page 146.) As one might expect, when Pasteur was proved right everybody believed that Pasteur was *totally* right, that *all* diseases were caused by germs. But some diseases (e.g., cancer) can be considered biochemical.

And the germ theory still has its detractors. In *The Blood Poisoners* (1965) Lionel Dole tells us that vaccination doesn't work, and that the only reason we all think it does is that the vast chemical combines, which have a vested interest in orthodox medicine, bring pressure to bear on the media.

Of course, physicians would have an enormous advantage if they could somehow confront at the molecular level the ailments they deal with, to take them on *mano a mano*, as it were. This is what the Canadian–Swiss anthropologist Jeremy Narby claims, in his book *The Cosmic Serpent: DNA and the Origins of Knowledge* (1998), the shamans of the Upper Amazonian rainforest are able to do by drinking a hallucinogenic tea called *ayahuasca*. Narby tried some of this tea, and discovered himself in telepathic conversation with twin serpent-like beings who were pure DNA – they were the famous DNA double helix, in fact, but also the twin or coiled serpents found in mythological art from all over the world (as on the staff of Aesculapius, Roman god of medicine). Narby concluded that they represented the consciousness of DNA. The shamans are able to cure people because, through drinking *ayahuasca* and invoking guided hallucination, they are able to get their consciousnesses down to the molecular level and thereby deal directly with "entities" (DNA serpents) and "darts" (which Narby equates with viruses). Needless to say, the scientific community has not taken this information very seriously, although three microbiologists Narby took to Peru to try the experience for themselves were apparently more impressed. A 45-minute documentary movie was made by Glenn Switkes of Narby's and the three microbiologists' adventure: *Night of the Liana* (2002), which has enjoyed a cult success.

An extraordinarily large number of theories concern constipation and, although most regard it as a Bad Thing, to be avoided by use of colonic irrigation (see page 290) or similar unsavoury means, this is not universally the case. From Benjamin Walker's *Encyclopedia of Metaphysical Medicine* (1978): "Although constipation was never deliberately employed as a technique in mysticism, it may have made its contribution to the visionary states of ascetics living on a sparse and inadequate diet." In fact, in bygone times constipation was almost exclusively a rich man's disease. The poor had as the basis of their diet extremely coarse bread, and ate very little meat; the rich, on the other hand, could afford luxury foods – and paid the penalty.

Some late-19th-century doctors carried out "orificial surgery" – i.e., surgery to enlarge the anus – because they believed that pressure within the rectum interfered with the proper development of adolescents: just like rock'n'roll in later decades, too much constipation might make them promiscuous. These surgeons were, of course, subscribing to the theory of focal sepsis (see page 290) – although it is tempting to suggest they subconsciously remembered the mediaeval belief that the bowel was the abode of demons: thus regular bowel movements were essential in case of a build-up of internal evil. For this reason, strong laxatives were employed in some exorcisms.

In more recent years naturopathists, notably Barbara Cartland, have cited constipation as the cause of every possible sort of disease, suggesting that evacuation every 12 hours is essential to good health. More extreme, perhaps, was Harry

Benjamin (1885–1986), whose *Better Sight Without Glasses* (1929) claimed that constipation caused eye cataracts and rheumatism. Both theories are marred by the fact that the unease caused by constipation is due not to poisoning by the retained matter but to the sheer physical discomfort of having a crammed bowel.

Since there are still so many myths current concerning this common complaint, it is worth noting that doctors report that many people who live quite happy, normal lives evacuate only two or three times a week; some people may defecate far more rarely than this without apparent ill-effect, their major discomfort occurring only during the eventual titanic evacuation.

General Semantics should be mentioned as a therapy. Its underlying theory is not total nonsense, although not original: in order to gain a fuller, more accurate worldview we should question deductions made by straightforward "Aristotelian" logic. That is, we tend to think of things as being either true or false, like a digital computer does, when in fact there are an infinite number of shades in between. Similarly, even a "definite" word like "house" is in truth indefinite – houses come in all shapes and sizes. Moreover, even a single house changes over time, so we should classify it in our minds not just as "house 1" (or "house 256", or whatever) but also as "house 1981", "house 1993", "house 2006", and so forth.

All of this is fair enough, and possibly of psychotherapeutic help to people who learn to think in General-Semantics fashion. But it is the range of ailments which General Semantics is supposed to cure which staggers – like so many other fringe therapies, it soon became regarded by its supporters as a general cure-all. Martin Gardner (b1914), in *Fads and Fallacies* (1957), told of a dentist who claimed it helped his patients: because they became more emotionally stable, their mouths became less acid.

One of most bizarre *faux* mental diseases ever to enter the textbooks was drapetomania, in vogue in the US for a few decades in the middle of the 19th century. Drapetomania was a mental ailment supposedly suffered by Black slaves who tried to flee the Southern plantations or otherwise displayed strong discontent with their bondage. The psychiatric reasoning was based in the myth that for Blacks, as an inferior human species, the norm was a condition of servitude; thus any Black who sought a different condition must necessarily be psychologically abnormal.

Faith healing is beyond the scope of this book, but a couple of psychic therapies are worth noting. In the 1960s Bryn Jones developed "somatography", a way of healing ailments by massage. But surely this is just the same as osteopathy or chiropractic (see page 300)? Well, in a way, yes – except that the massage is applied not to the patient's body but to her or his aura.

A similar technique is "therapeutic touch". Despite the name, the therapist does *not* touch the patient, but instead runs his or her hands over the patient's "life-field" – a few centimetres from the body. It would seem the successes of this therapy are due to the practitioner's movements and patter assisting the patient to visualize bodily malfunctions, and to believe that it is in her or his power to cure these by effort of will. The therapy is, therefore, a sort of biofeedback without the feedback.

In 2004 the *British Medical Journal* released online a patients' guide to therapies, *BMJ Best Treatments*, edited by Luisa Dillner, which admitted that for many ailments the best possible therapy is no therapy at all, while often there is no scientific underpinning for or empirical evidence in support of the most-used therapies. Among the revelations:

- The removal of young children's adenoids, thought to improve breathing and prevent ear infections, is probably a waste of time. Most of the problems will clear up with time.
- While anorexia is commonly treated with a combi-

nation of drugs and therapy, there is no compelling evidence that any of the treatments in use are actually effective.

- There is no treatment for anxiety that's known to be effective, although cognitive therapy (akin to counselling or psychotherapy) and antidepressants may help. However, the guide cautions that antidepressants should be used for short periods only.
- For breast cancer the commonly used technique of mastectomy – removal of the entire breast – has been shown to be no more effective than removal of only the tumour.
- There is no evidence that the use of grommets (small tubes placed into children's eardrums) as a cure for glue ear is effective. The best idea is to leave the ears alone and wait for the child to grow out of the condition.
- People suffering from prostate cancer who merely "wait watchfully" have the same life-expectancy as those who undergo the commonly prescribed surgery, radiotherapy and hormone treatment.
- Extraction of the tonsils as a means of deterring sore throats is probably pointless: there's no evidence that the surgery helps at all. As for preventing ear infections, the other main reason given for tonsillectomies, antibiotics work as well or better.
- In cases of impacted wisdom teeth, where extraction is the customary prescription, this may do more harm than good. The best advice, if they're not causing problems, is to leave them alone.

Of course, the classic place to go to be cured is Lourdes, in France, where a staggering three million of the faithful arrive each year in hopes of a miracle. The popular conception of Lourdes is that the cures occur pretty frequently, but no. In order to try to fend off charlatanry, the officials at Lourdes exercise a strict control system. As just one example of the kind of control that quacks of all stripes, not just faith healers, so often ignore, they conduct a thorough check beforehand to make sure the individual is actually suffering from an ailment. Once that has been established, the process is far from over, the final control being that the cure must be authenticated as having lasted for a period of years. Only then will the case go down in the books as a genuine miracle. Perhaps because of these stringencies, a cure is recognized only about once every seven years, on average – that is, about one cure per every 21 million visitors.

Focal Sepsis

Also known as focal infection and, in dentistry, oral sepsis, this belief was responsible, up until about WWII (with some survival of the practice even today), for untold numbers of ghastly mutilations being performed in the name of cure. The underlying claim was that mental and physical illness comes as a result of toxins absorbed into the bloodstream from a focus of sepsis (i.e., a clump of bacteria) within the body: the obvious solution to the ailment was to excise the offending organ. Various potential sites for these foci were postulated. Thus chronic constipation (the bowels as focus of the sepsis) resulted in many patients being subjected to repeated colonic irrigations and/or surgery to enlarge the anus. Often *all* of a patient's teeth would be removed on the flimsiest evidence of dental caries in just one. Appendectomy and colisectomy were routine, even if there were no clinical justification for such operations. The nasal sinuses, too, were regarded as a prime site for chronic sepsis: patients were often subjected to an almost routine submucous resection or to more drastic operations in order to improve drainage.

In the middle of the 19th century the germ theory (see page 285) reached the UK and a few decades later it arrived in the US. For a while germs were invoked as the cause of every possible ailment. In particular, attention was paid to the germs in the gut, and colonic purging, using powerful laxatives or the painful process of colonic irrigation, was prescribed for maladies that included peptic ulcer, gastric cancer, endocarditis, arthritis and even straightforward stupidity! Colonic irrigation was a process even grimmer than it sounds: it involved pushing a tube into the rectum and forcefully flushing large quantities of water through it. What such treatments primarily achieved was flushing out the colon's resident bacterial flora, thereby removing much of that area's resistance to further infection.

It was perhaps in dentistry rather than mental health that the doctrine of focal sepsis played its biggest part. The dental variant of the concept, oral sepsis, was first to our knowledge recorded by Hippocrates (*c*450–*c*370BC), who noted that a case of arthritis had apparently been cured by a patient's having had a tooth pulled out. Early in the 19th century the US physician Benjamin Rush (1746–1813) – one of the signatories to the Declaration of Independence – concurred that arthritis could be cured by dental extraction. But the craze for focal-sepsis-oriented treatments really began with the UK physician William Hunter (1861–1937), with a 1900 paper in the *British Medical Journal* blaming all sorts of ailments on poor oral hygiene and the increasing practice by dentists of attempting to save teeth rather than extract them, and more especially with a hugely influential lecture he delivered in 1911 at McGill University, Montreal:

> No man has more reason than I to admire the sheer ingenuity and the mechanical skill constantly displayed by the dental surgeon. And no one has had more reason to appreciate the ghastly tragedies of oral sepsis which his misplaced ingenuity so often carries in its train. Gold fillings, crowns and bridges, fixed dentures, built on and about diseased tooth roots form a veritable mausoleum over a mass of sepsis to which there is no parallel in the whole realm of medicine and surgery. A perfect gold trap of sepsis of which the patient is the proud owner and no persuasion will induce him to part with it, for it cost him much money and covers his black and decayed teeth. The worst cases of anemia, gastritis, obscure fever, nervous disturbances of all kinds from mental depression to actual lesions of the [spinal] cord, chronic rheumatic infections, kidney diseases, all those owe their origin to, or are gravely complicated by the oral sepsis produced by

> these gold traps of sepsis. Time and again I have traced the very first onset of the whole trouble to the period within a month or two of their insertion.

One might expect from this that Hunter would have cited in his lecture a rash of pertinent cases as supportive evidence for his claims, but in fact he offered just one, and that an example of where a dentist had given the patient a stupid instruction.

In 1912 the US physician Frank Billings (1854–1932) established the focal-sepsis theory in more general medicine, not just citing plentiful case histories but also claiming to have himself effected numerous cures of various maladies through pulling teeth and extracting tonsils. He also reported having taken microorganisms from arthritic patients and injecting them into rabbits, who in due course likewise developed arthritis. His pupil E.C. Rosenow (1875–1966) expanded the theory further, introducing two new concepts: "elective localization" and "transmutation". The first of these held that specific germs gravitated toward specific parts of the body. The second maintained that, once there, they could transmute into different germs. This latter was an especially effective means of explaining why other researchers had difficulties replicating Hunter's and Rosenow's experiments: they were looking for the original germ that had been injected rather than the germ into which it had transmuted; since they were bound to find germs of *some* kind . . . well, that was just additional proof that germs did indeed transmute, wasn't it? Matters weren't helped by the fact that Rosenow, when injecting his lab animals with microorganisms, typically did so intravenously, in such quantity and with such virulent germs that practically no part of the unfortunate creature was left unaffected.

In the 1920s another US physician, Weston A. Price (1870–1948), published a series of results demonstrating that, through extracting the teeth of rabbits infected with all kinds of diseases, he had effected cures or gained significant improvements in the animals' condition. It was his conclusion, therefore, that wherever possible infected teeth in humans should be extracted rather than any attempt being made to save them.

Many eminent physicians accepted what Hunter, Rosenow and Price advocated, and US medicine embarked on a massacre of the nation's teeth, tonsils and adenoids. Soon all kinds of other parts of the body were being surgically removed, even limbs amputated. The rich suffered the attentions of the surgeons far more often than did the poor, by a ratio of about two to one in the US and about three to one in the UK.

But there were dissenting voices both within the medical profession and outwith it. As early as 1926, in the US, Nicholas Kopeloff was observing in *Why Infections? In Teeth and Other Organs*: "If this craze of violent removal goes on, it will come to pass that we will have a gutless, glandless, toothless – and I am not so sure that we may not have, thanks to false psychology and surgery, witless race." More and more, papers began appearing in the scientific journals that questioned the theoretical bases of the focal-sepsis principle, complete with extensive series of experimental results that showed pretty clearly that all the surgeries and extractions were doing very little good – and indeed in many instances were increasing rather than decreasing infection rates and/or exacerbating the original condition, quite aside from the disadvantages of the patient henceforth lacking this, that or the other useful bodily part.

A paper published in 1940 in the *Journal of the American Medical Association* by H.A. Riemann and W.P. Havens marked the real turning of the tide: they pointed out that the focal-sepsis theory was unproven, that there was no evidence that the excisions or extractions had any beneficial effect and plenty of known instances where the operations had done harm, and that there were numerous instances where the diseased sites that were the supposed foci of infection had been cured by clearing up, through diet or other means, the systemic ailment whose source they were supposed to have been. Even so, the practice in fact continued for decades afterwards of quite unnecessarily removing children's tonsils and adenoids, sometimes in the mistaken belief that these were the source of other maladies, and sometimes as a matter of course in supposed prevention of future ailments. The practice of removing the uterus for reasons unrelated to the uterus itself likewise continued: a 1970s survey done in the US showed that about 30% of hysterectomies were "probably unnecessary".

Official historians of the medical sciences have chosen largely to ignore the focal-sepsis craze of the 20th century's first half (and later). This is possibly through sheer embarrassment, a denial of some of the horrors that were perpetrated in the name of an almost entirely unsupported medical dogma. Nowhere were the horrors greater than at the Trenton State Hospital (New Jersey State Hospital for the Insane) under the aegis of Henry Cotton (1876–1933), the hospital's Superintendent from 1907 to 1930 and the subject of the book *Madhouse* (2005) by Andrew Scull.

In 1913 the germ responsible for syphilis was isolated and discovered to be the cause also of a deadly mental disorder common at the time, called general paralysis of the insane (GPI). If this particular microorganism could produce one form of insanity, might not other mental illnesses likewise be caused by infections? It was Cotton who put such notions together with the increasingly popular concept of focal sepsis as well as one that was becoming emergent, mental hygiene, the notion that steps could be taken to *prevent* mental illness before it struck. The dental aspects of focal sepsis were then to the fore, so Cotton began removing teeth from his patients, initially just teeth that showed some sign, however slight, of infection, but in due course the entire mouthful, just to be on the safe side. But of course that was to look only at the small picture. Those infected teeth must surely have spread toxins into the patients' saliva, which of course the patients would have swallowed, taking the toxins into the digestive system. The toxins probably ended up in the colon, so colonic irrigation was prescribed. But maybe that wasn't doing a thorough enough job: partial colectomies became the order of the day, and then total colonic excisions. With hindsight one can see the progression of Cotton's thinking as obsessive; at the time far too few of his colleagues did. He drained sinuses, removed tonsils and testicles, gall bladders and cervixes, stomachs and spleens . . . on and on the surgical operations went.

There were, though, some sceptics. One of these was Cotton's mentor, Adolf Meyer (1866–1950) of Johns Hopkins Medical School, who had not only trained Cotton but been influential in getting him the post at Trenton. Meyer, with Cotton's eventual concurrence, in 1924 sent in an assistant, Phyllis Greenacre (1896–1989), to double-check the 85% cure rate that Cotton was claiming. Greenacre, and later Emil Frankel in a second report, found the picture to be entirely different from the one Cotton was painting: at best the recovery rate was 23%, and the mortality rate was horrifyingly high – among those patients who had suffered a total colectomy, it was about 45% (138 of 309 patients). The mortality rate for those who'd undergone partial colectomies was somewhat lower, but the percentage of those patients still in mental hospital, institutions for the feeble-minded or indeed prison was correspondingly higher: although they'd survived the operation, in no sense could they be said to have been cured by it. Perhaps unconsciously, Cotton was manipulating his own data to present the picture of a high success rate by blaming other factors for continuing illness or death – or even, in instances of continuing mental illness, by believing that the original surgery had not been radical enough so the patient should go back for more!

Astoundingly, for reasons connected with psychiatric politics and the desire to avoid a scandal, Meyer suppressed Greenacre's results (Frankel's later results were out of his domain). And so the butchery at Trenton continued. In the end, it was more New Jersey's notoriously dirty politics than anything else that got Cotton removed from his post. His successors did much to cover up the nightmare that had gone on before, and consequently the full story did not emerge into the spotlight until, largely thanks to Scull's work, the beginning of the new century.

The ironic thing about all this is that focal sepsis is not

entirely a false idea. In certain very specific instances a bacterial or viral infection in one localized part of the body can indeed cause a system-wide ailment. However, the bad name given to focal sepsis by the overenthusiasm for it as an explanation for such a huge range of ailments during the first half of the 20th century has made medical scientists naturally reluctant to invoke it when trying to find causes for disease today.

HOMEOPATHY

Homoeopathy was the brainchild of the German physician Christian Friedrich Samuel Hahnemann, whose major work, *The Organon*, was published in 1810. The principle of homoeopathy is to give the patient mild doses of whatever might produce the symptoms from which he is already suffering. This is according to the theory that symptoms are not a *result* of the disease but are manifestations of the body's way of *coping with* the disease. Thus the homoeopathist is assisting Nature by boosting the symptoms.

"Mild doses" are indeed called for. Homoeopathic doses should always be very, very dilute. Patients may be given as little as one decillionth of a grain; a decillionth is 1 divided by 100 million million million. With doses as small as this – the smaller the better, in most cases – it is hardly surprising to discover that in certain cases Hahnemann recommended that the solution be so diluted that *not a single molecule* of the curative drug reach the patient's lips.

A late-19th-century homoeopathist named Wilhelm Heinrich Schüssler (1821–1898) produced a branch of homoeopathy which he called "biochemistry" (not to be confused with real biochemistry). In an idea reminiscent of the humours and of one of Paracelsus's notions (see page 266), he said our tissues are essentially made up of 12 "salts"; the basis of his therapy was that an illness is the result of a deficiency in one or more of these "salts", which deficiency can of course be amended. His modern followers have increased the number of salts involved to about 40.

Vervain plant from 13th century English manuscript

The system of Bach Flower Remedies is a variant of homeopathy first developed in the early 1930s by the UK physician Dr Edward Bach (1886–1936), who based his therapy system on the perfectly reasonable notion that at least some physical ailments are caused by emotional imbalance:

> Disease is in essence the result of conflict between Soul and Mind, and will never be eradicated except by spiritual and mental effort. Such efforts, if properly made with understanding . . . can cure and prevent disease by removing those basic factors which are its primary cause. (Edward Bach, *Heal Thyself: An Explanation of the Real Cause and Cure of Disease*, 1931)

Bach noticed that animals frequently lick dew from leaves and flower petals, and came to the conclusion the animals were in effect self-medicating. Research led him to devise a total of 38 flower infusions (as with homeopathy proper, more have been added by subsequent researchers), which were created to homeopathic dilutions by exposing to sunlight containers of water with flowers floating on the surface. It was Bach's contention that the efficacy of his remedies came from the psychic energy of the flowers.

Although Bach died young, the institution that he founded, the Dr Edward Bach Centre, still flourishes in Brightwell-cum-Sotwell, Oxfordshire, UK.

NATUROPATHY

A fringe medical school (or, really, cluster of schools), naturopathy proposes that illness can be treated by purely natural means – water, unprocessed foods, diets, exercise – and warns against such dangerous practices as taking aspirin or going to see your doctor. It can trace its antecedents back to Hippocrates (*c*450–*c*370BC) and his *vis medicatrix naturae* (the healing power of Nature). Another great figure in naturopathic history was Thomas Sydenham (1624–1689), the "English Hippocrates", who proposed that illness itself was a form of therapy – in many ways a perceptive idea, since the complex of symptoms we call a "cold" is indeed the body's attempts to cure itself. But extended to, say, smallpox the idea was a little dangerous. The great pioneer in the 20th century was the US Seventh Day Adventist John H. Kellogg (1852–1943), who invented the cornflake, which his younger brother, Will Keith Kellogg (1860–1951), marketed.

Georgei Ivanovich Gurdjieff (*c*1873–1949) seems to have invented a bizarre amalgam of naturopathy and drug-oriented medicine – he appears to have enjoyed disagreeing with the medical establishment as a matter of principle. One young woman who came to his Institute for the Harmonious Development of Man, at Fontainebleau, was suffering from tuberculosis. Nevertheless, Gurdjieff exhorted her to sleep in the draughty loft above the barn. At first, this treatment seemed to be working, but a week later she was dead, aged 35. She was the writer Katherine Mansfield (1888–1923).

A surprising number of people over the last century or so have believed that fasting is of medical value – as either prevention or cure. Naturally, it should be used for the usual dreary catalogue of serious diseases – cancer, diabetes, etc. Of course, there is little doubt that most of us eat too much, so that an occasional pause for a day or two won't hurt and might do some good. However, prolonged fasting is obviously harmful and, equally obviously, if the patient is suffering from any of a number of serious diseases, can be fatal.

One school of naturopaths maintains stoutly that germs don't cause disease – diseases cause germs. What we think are infecting bacteria are in fact deformed cells produced by our bodies as a result of the disease. So much for bacteriological warfare.

Politics and naturopathy have often flirted. Whenever traditionalist or nationalistic leaders feel threatened by science, they feel threatened, too, by scientific medicine. Thus the Ayatollah Khomeini hits out at the traitors who have "encouraged a handful of inexperienced young men to study this accursed European medicine", adding that illnesses "such as typhus, typhoid fever and the like are curable only by traditional remedies". And the Nazis were, as might be expected, naturopathy fans, too. Dr Otoman Zar-Adusht Ha'nish (Otto Hanisch; 1854–1936) founded the Mazdaznan, a society which still exists. This believes in naturopathy with a difference: only Aryans benefit from the treatments.

OSTEOPATHY AND CHIROPRACTIC

Osteopathy was founded in 1874 by the Virginia physician Andrew Taylor Still (1828–1917), who was impelled to seek a better form of medicine after an epidemic of viral meningitis killed three of his children despite the fact that he, their own father, was a trained doctor. In his re-evaluation of the medical science of his day he (correctly) fixed upon overfrequent amputation and the overuse of drugs as major flaws, and instead adopted a more holistic approach: the body was, he said, a single entity, so that it was impossible for one part of it to be in disorder without other parts malfunctioning as well. He therefore devised a new system of therapy whose underlying hypothesis maintained that all ill-health is as a result of minute dislocations of the vertebrae, which bones press on nerves or blood vessels to cause "subluxations". These are stoppages in the passage of blood or (presumably) some analogous nerve fluid; and they obviously result in the relevant fluid becoming stagnant, so counteracting the effects of the curative constituents the fluid would normally be transporting around the body. The remedy is spinal massage. He called his therapeutic system osteopathy after the Greek word *osteon*, meaning "bone".

Still set up what would become his first osteopathic medical school in Kirksville, Missouri, in 1892, having failed to persuade any traditional centres to let him open a school under their auspices. Over the course of years he made some outrageous claims for his creation. He said that in one small US town he had in a single day reset no fewer than 17 dislocated hips. We do not know why so many people had dislocated their hips all of a sudden. Osteopathy, Still seems to have believed, could cure every disease known to humanity – including baldness! In *The New Apocrypha* (1974) John Sladek tells us that Still claimed to have grown 7.5cm of hair on the head of a previously bald man during the course of one week. Since Still's time, osteopaths have begun to claim that their therapy is useful in the treatment of mental illnesses, too – which it may in fact be, since massage of any form can have a soothing effect that might well be therapeutic in specific instances.

An extension of Still's original theory was cranial osteopathy. The cranium is formed of 22 bones that fuse together in very early infancy (plus the mandible of the jaw and the bones of the ear). However, these 22 bones may not fuse together in precisely the correct alignment; moreover, the alignment may be disrupted in later life by physical trauma – whiplash, a blow, or even the stress of having a tooth out. Particularly vulnerable is the hinge of the jaw, which can relatively easily be slightly dislocated. According to the US physician William Garner Sutherland (1873–1954), who for a time around the turn of the 19th and 20th centuries was a student of Still's, the cranium was capable of a motion analogous to respiration; furthermore, the cranial bones were not in fact fused together but were in a constant state of tiny rhythmic movements relative to each other along the lines of the flexible sutures between them. Sutherland went on to combine this insight with his osteopathic knowledge.

The heart of his therapy was the concept of the primary respiratory mechanism, or PRM. This is not to be confused with ordinary respiration, but is, rather, a fundamental system of the body in the same way that, say, the lymphatic system is essential to our wellbeing.

The functioning of the PRM depends upon the flexibility of the brain and spinal cord, which pulsate in a steady rhythm: during the "exhalation" phase the brain and skull get taller and thinner, while during the "inhalation" phase they get shorter and broader. These movements affect the shapes of the cavities within and surrounding the brain and spinal cord, thereby imparting pressure waves to the cerebrospinal fluid, distributing it – so the hypothesis claims – to the rest of the body via the spinal nerve sheaths (a movement that has not in fact been observed). The brain is surrounded by membranes (the meninges), of which the outermost, largest and toughest is the dura mater. When the head receives a trauma, these membranes, the dura mater in particular, can be pulled out of kilter, and this affects the "fulcrum" around which the cerebrospinal fluid rhythmically moves; the membranes can become so distorted as to force the cranial bones out of alignment with each other.

A further hypothesis here is that bone is not a solid material but a superviscous liquid (much like glass is a superviscous liquid even though we think of it as a solid). A traumatic blow can of course shatter bone, but a lesser intensity can, as it were, stun the bone into rigidity – i.e., deprive it of its fluidity and flexibility. The spine is connected to the skull via the dural membranes of the spinal cord, allowing us a flexibility of head movement but also meaning that traumas experienced by the spine/spinal cord can easily be transmitted to the head, and vice versa. The entire system – cranial bones, spinal bones, membranes, brain, spinal cord and cerebrospinal fluid – is thus a single entity, and imbalance in any one part of it inevitably leads to functional disturbance elsewhere.

The therapist obviously cannot, without surgery, directly affect most of the system's components, the exceptions obviously being the spine – the province of the traditional osteopath – and the cranial bones, which is where the cranial osteopath comes in, manipulating the skull much as a traditional osteopath manipulates the vertebrae.

The above discussion should not be taken to mean that the modern osteopath need necessarily be a quack, and her or his treatment valueless. This is because the modern practitioner will generally make use also of all the paraphernalia of modern medicine, and will massage not only the spine but also any other area of the body. As an *adjunct*, osteopathy seems at worst harmless and at best, as in cases of uncomfortable but subcritical spinal misalignment, beneficial.

Chiropractic is much the same as osteopathy, only with an even less rational theoretical underpinning. Alternatively, it may be thought of as a form of zone therapy (see page 308), one in which the control points are ranged along the length of the spine rather than in the fingers. The practitioner diagnoses your disease, presses the appropriate button on your spine, and you will be helped or even cured.

Daniel David Palmer (1845–1913) instituted the practice of chiropractic in 1895 when, according to his own account, he cured a man's deafness through manipulating a bump on the spine; this had the effect, said Palmer, of easing pressure on the nerve that ran to the ear – odd, since the nerve running from the brain to the ear doesn't go anywhere near the spine. Palmer was jailed in 1906 for practising medicine without a licence, but this didn't deter him. And it hasn't deterred chiropractors since: a survey of chiropractic colleges done in the late 1960s revealed that fewer than half of the teachers there had themselves graduated from college, and far fewer still had degrees relevant to the subjects they were teaching. More recently the colleges are said to have cleaned up their act a bit.

One chiropractic habit that has drawn particular criticism from the medical profession is that of requiring extensive X-rays of the spinal column. This is a dangerous procedure. The sexual organs, for example, receive a dose of radiation about 1000 times higher during a spinal X-ray than they do during a chest X-ray.

COLOUR THERAPY (CHROMOTHERAPY)

The belief that coloured lights can be used in therapy is an old one, based on the emotions which we associate with the different colours – martial red and "seeing red", cool blue and "the blues". The elevation of the technique to the status of full-blown pseudotherapy was a product of the 20th century. One well known form was the Spectro-Chrome Therapy of Colonel Dinshah Ghadiali. Once you had determined your ailment, all you needed to do was slide the correctly coloured pane of glass into the Spectro-Chrome machine which you had hired, switch on the bright lights behind the glass, lean back, and bask in the healing rays of coloured light. Because that might seem too simple, you had also to undergo dietary restrictions – essentially, you had to give up everything you enjoyed eating or drinking. Spectro-Chrome Therapy claimed to cure even such diseases as diabetes and appendicitis.

The strangest colour-therapy variant has probably been that of William Estep (1920–2000), who by shining coloured lights on water could turn it, Christ-like, into medicine.

BIORHYTHMS

The notion of biorhythms enjoyed a surge of popularity during the 1970s and 1980s, but is not much heard about today. It owes its origins to the ideas of the Berlin ENT surgeon Wilhelm Fliess (1858–1928). For a while Fliess was associated with Freud, and during this time he came up with the idea of vital periodicity: vital processes are dependent upon a cycle, this cycle being of duration 23 days in men and 28 days in women. (This leads to the depressing conclusion that a married couple will together be at their peak of vitality only every 644 days.) Further cycles had durations combining the 23 and 28 basic values – for example, 51 (23 + 28) days. Later, during the 1920s, the Austrian engineer/mathematician Alfred Teltscher (of whom little further is known – see below) proposed a third, 33-day cycle, concerned with the intellect, to be superimposed upon the other two; supposedly Teltscher had deduced this 33-day cycle from observation of the students in his classes. Thus any individual was subject to a 23-day cycle involving such supposedly masculine attributes as physical strength, a 28-day cycle involving supposedly feminine attributes like intuition, creativity and sensitivity, and a 33-day cycle involving such non-gender-specific attributes as reasoning and ambition. Viennese psychologist Hermann Swoboda (1873–1963) added the notion that all three cycles began at birth, so that it was relatively easy, knowing any individual's birthdate, to calculate the state of their vitality.

Fliess himself was adamant that the female 28-day cycle was not to be equated with the menstrual cycle: the only connection between them, he maintained, was that they were of similar evolutionary origin. Fair enough, but in that case it's legitimate to ask what are the evolutionary bases of the 23- and 33-day cycles. Since biomedical researches showed no trace whatsoever of the three cycles, biorhythms eventually faded out of the popular consciousness . . . leaving just one mystery: *Who was Alfred Teltscher?* He is frequently described in the literature as belonging to the University of Innsbruck, but this seems to be in reference to a Friedrich Teltscher, who was there and who published in 1918 a chapbook on the electrical properties of mercury and who thus approximately fits the description, but only approximately. One cannot say with certainty that Alfred Teltscher was an invention – perhaps by Swoboda (who among other things had a habit of claiming Fliess's "discovery" of the first two cycles as his own) – but the situation does seem, to say the least, odd.

Eye Exercises

Biologists are wrong about how the eye focuses. They think the lens of the eye becomes fatter when you focus on nearby objects (the fatter the lens, the shorter its focal length), whereas what *really* happens is that the lens moves backwards and forwards to focus the images of objects at different distances onto the retina – much as a camera is focused by moving the lens backwards and forwards. Or so said Dr William Bates (1860–1931) and his followers. Actually, this is exactly the way in which the eyes of some animals – e.g., fishes – do focus; it's just that the human isn't among those animals.

In the camera-type-focusing eye, muscles squeeze the eyeball as a whole, thereby increasing the distance from lens to retina, or pull outwards to decrease the distance. According to Bates, then, the cure for sight defects was not the use of spectacles but frequent eye exercise designed to relax the external muscles employed in the squeezing, or otherwise, of the eyeballs. He and various followers devised numerous variants of such exercises, some of them dangerous – such as Bates's recommendation that you strengthen your eyes by staring at the Sun.

Batesian eye exercises are still popular, alas. In a book of mine published in 1981 I made a remark to the effect that I doubted if anyone following these exercises would be reading the book. In all the mail I received following the book's publication, not one reader challenged me on this statement.

A completely different school of pseudomedical thought concerned with the eyes is iridiagnosis. This has it that physicians can diagnose their patients' illnesses simply by staring them straight in the eyes. A pseudomedical diagnostic technique invented towards the end of the 19th century, iridiagnosis is based on the notion that the iris is divided into some 40 zones, each corresponding to an area of the body. Close observation of the iris can therefore reveal which part of the patient is ailing.

ACUPUNCTURE

Science as yet keeps an open book – just open – on the practice of acupuncture, which does seem to be efficacious in some instances where the results cannot be simply explained away in terms of the placebo effect. The underlying theory used to explain the effects of acupuncture is, however, in general roundly rejected by occidental science, drawing as it does far more on metaphysics than on reality. One hypothesis advanced in hope of explaining acupuncture's attested beneficial effects is that the pricking of the body stimulates the generation of endorphins, which are the body's own natural painkillers. Of course, simply because pain has been reduced doesn't mean the acupuncture has had any effect on the illness which created that pain: much more research will have to be done to see if acupuncture genuinely aids curing or if it's simply a handy pain-reliever.

The metaphysical rationale offered for acupuncture's presumed benefits relates to the ancient Chinese concept of yin and yang, the two principles said to govern the flow of the Universe. The healthy body contains a perfect balance of yin and yang, in this context yin being interpreted as (in essence) chronic conditions and yang as (again simplistically) acute ones: a condition like yuppy flu (chronic fatigue syndrome) would thus indicate an excess of yin, while a blinding migraine would be symptomatic of excess yang. The lifeforce of the Universe, the ancient Chinese believed, flows through the body along certain precisely defined paths, known today in the West as meridians; certain points (nodes) on these meridians link to specific bodily organs. Centuries ago Chinese physicians mapped out the meridians and nodes. The acupuncturist believes that, by sticking needles into the body at the nodes and agitating them (either through physical manipulation or, often these days, by putting a low-voltage electric current through the needles), s/he can adjust the flow of the lifeforce in such a way as to bring the yin/yang equation back into balance.

Originating like acupuncture in China, and based upon the same philosophical/metaphysical premises, acupressure involves massaging various predetermined areas of the body – in effect, it's acupuncture without the needles. From China the practice went to Japan, where it thrived as a practice until the 19th century, at which point it was banned as being merely massage under another guise – i.e., part of the entertainment industry rather than of therapeutic nature. That law was repealed in 1955, and the practice became popular once more, soon spreading from Japan to the West.

A variation of acupressure is acu-yoga, which as one might expect from the name brings yoga into the mix. Here the individual, rather than a therapist or masseuse, is responsible for the application of the pressure to the appropriate points while adopting prescribed yoga positions, with the floor and/or wall being brought into play so that pressure can be applied to those parts of the body the hand or foot can't reach.

Zone therapy seems to be based on principles similar to

those of acupuncture, with the main differences being that it's far simpler and that it certainly doesn't work. The body is divided into 10 zones, to each of which corresponds a finger and a toe (sometimes the tongue also plays a part). Bodily pains can be eased and illness cured by applying pressure to the right combination of fingers and toes – and the tongue, if necessary.

PHRENOLOGY

Phrenology was devised – although he did not himself use the term – by Franz Josef Gall (1758–1828) before 1800, and enjoyed a tremendous vogue during the 19th century; today it has few practitioners. In its heyday its supporters included Alfred Russel Wallace (1823–1913) and Walt Whitman (1819–1891) – whose devout enthusiasm was based on the knowledge that his own skull-bumps showed him to be well endowed in every conceivable respect.

Gall's system was based on the following hypotheses:

- The brain is the seat of the mind.
- The mind is not a unitary thing but is made up of a number of distinct faculties.
- Each of these faculties has a distinct centre – an "organ" – in the brain, the shapes of these organs, taken together, determining the overall shape of the brain.
- As a rule of thumb, the bigger any organ is by comparison with other organs of similar function, the more powerful it is.
- Since the skull takes its shape from the brain, the surface of the skull can be "read" as a map of the brain and hence of the various "organs" – and hence again as a measure of the individual's faculties.

Gall divided the brain up into 27 zones, each of which corresponded to a human quality or characteristic. Naturally, a hyperintelligent person (say) would have a larger volume of brain in the intelligence zone than the rest of us, and Gall

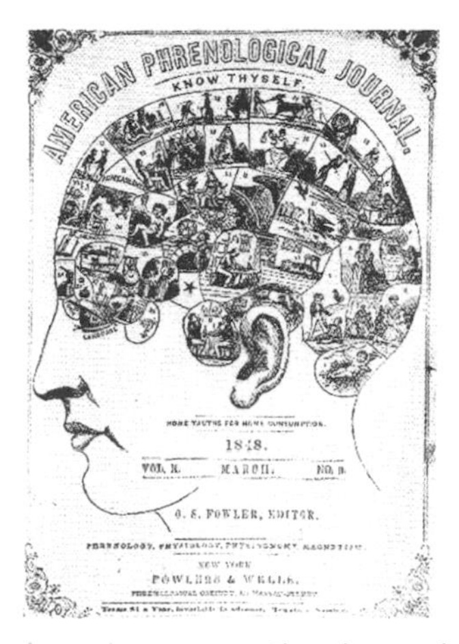

believed this would manifest itself as a bump in the appropriate region of the skull.

Unfortunately, while there *is* localization within the brain of the sensory centres and some of the mental functions, *characteristics* are not localized there. Moreover, the shape of an individual's brain does not directly affect the shape of his skull – even on the inside – since nowhere does the brain actually touch the bone. In evolutionary terms there may be a change over time in cranial shape to accommodate developments in the brain, but this is irrelevant in the context of an individual skull.

Gall and his important convert Johann Gaspar Spurzheim (1776–1832) promoted their new science eagerly, but it wasn't really until 1815, when the *Edinburgh Review* published a blistering attack on the practice, that phrenology found its way into the popular consciousness. Spurzheim published a vigorous defence of phrenology against the *Edinburgh Review* article, with the result that it was in Edinburgh, in 1820, that the first phrenological society appeared, and the Edinburgh lawyer George Combe (1788–1858) would remain an important figure in phrenology for decades; many further phrenological societies followed on both sides of the Atlantic. The name "phrenology" was coined in 1815 by Dr Thomas I.M. Forster (1789–1860).

During phrenology's heyday, although it was never other than under attack by mainstream medical scientists, it could be taken very seriously indeed by much of society. Some employers would demand a phrenological analysis before taking on a new employee to check he was sufficiently intelligent and of good moral character. Phrenology was applied also in the field of criminology, with the same disastrous results as

Lombroso's physiognomy. And, although most phrenologists were convinced that what they were practising was indeed science, phrenology attracted its share of Christian Creationists; Gall had included in his 27 zones organs such as those for veneration and wonder, and these "spiritual" zones seemed concordant with the notion of a Creator.

Phrenology had more or less died out in the UK by the early 1850s, but in the US it remained hugely popular, thanks in large part to the efforts of two brothers, Orson Squire Fowler (1809–1887) and Lorenzo Niles Fowler (1811–1896); they started their careers in New York in the 1830s and had made a small fortune from it. Lorenzo visited the UK in 1860 for a lecture tour that lasted a couple of years, in 1862 moving there permanently and in 1863 founding in London the Fowler Institute. He also began to sell phrenological paraphernalia on a wide scale, and the practice of "head reading" once more become all the rage among the English chattering classes. Most of the phrenological busts you still see around were Lorenzo Niles's products.

Unfortunately, with this new variant of phrenology came unpleasant racial overtones. It had always been the case that phrenological analysis was almost entirely subjective, whatever its practitioners might have thought: if a large bump for one faculty or another was emphatically contrary to the evidence of a person's actual character, there were various recognized ways of weaseling around this. If a phrenologist could read whatever he wanted to into the bumps of an individual's skull, the way was open for the phrenologist's prejudices to be amply confirmed – with an astonishing degree of consistency, in fact – by the skulls he "read". The racist connotations, although more noticeable in this later, revivalist phase of phrenology, were not unique to it. Earlier, in *The Constitution of Man* (1827), George Combe wrote concerning African skulls that they were

> overdeveloped in the Organs of Philoprogenitiveness and Concentrativeness (accounting for the African's alleged love of children and proclivity for sedentary occupations, respectively), and underdeveloped in Conscientiousness, Cautiousness, Identity and Reflection.

Phrenology withered on the vine again in the early years of the 20th century, probably simply because it became unfashionable rather than through any greater acceptance by the public of advances in genuine science. Even so, it continued to have reverberations in the field of physical anthropology, far too often once more for racist purposes. However, a few of its basic insights proved of value, especially the idea that mental faculties can have distinct seats in the brain: this is true of at least some faculties (others are distributed throughout the brain), and it is true also that, as the phrenologists guessed, these localized sites can increase in size with frequent and persistent use.

Physiognomy

Phrenology is really just a specialized version of physiognomy, the ancient belief that a person's character can be told from their outward appearance, most particularly their facial features. It was clearly current in the time of Aristotle – he toyed with the idea, but doesn't seem to have given it much time – and was popular through the Middle Ages. As a pseudo-science it was discussed by Giambattista della Porta (1535–1615) in Italy and Sir Thomas Browne (1605–1682) in England, before being revived in the 18th century by Johann Kaspar Lavater (1741–1801), a Swiss clergyman who based his ideas on their work. His essays were widely translated throughout Europe.

There are two basic variants of physiognomy. In the more absolute form, it is believed that there's no escaping the characteristics evidenced by your facial features: if the shape of your nose shows you're a criminal, then that's what you are. The less absurd variant says that the physical features are merely indicators of a character *tendency*: there is a good statistical correlation between nose shape and criminality (say), but far from all people with that nose shape are criminals and far from all criminals have that nose shape.

The use above of criminality as an example of the characteristics physiognomy could reveal is no accident, for the pseudoscience reached its apotheosis in the work of the Italian criminologist Cesare Lombroso (1835–1909). His criminological theory was based primarily on physiognomy, although it drew in also some of the ideas of eugenics. It was Lombroso's contention that criminals are born, not made: they are evolutionary throwbacks, atavisms. Since they could be regarded as having devolved to a more primitive evolutionary state of humankind, it was reasonable that this should be reflected by "atavistic stigmata" in their facial and other features. Lombroso embarked on a very extensive campaign of precisely measuring the physiological aspects of as many convicted criminals as he could in order to prove his thesis – which of course his statistical analysis did. Despite the seeming grimness of Lombroso's determinism, the motivation of these criminological studies was in fact surprisingly benign: an opponent of capital and other draconian punishments popular in his day, he was convinced that, once criminals had been identified through their features, the next step was to try to cure them of their criminality, or at least treat them in order to ameliorate it.

Later criminologists who adopted Lombroso's ideas tended to be significantly less humane, believing that physiognomy was a tool that could be used for the more efficient apprehension of criminals – so efficient, indeed, that with luck and skill you might catch them *before they had even committed a crime*.

Of related interest is the anthropometric system called bertillonage, devised by the Parisian police chief Alphonse Bertillon (1853–1914): the Parisian police, and soon those in many other countries, made systematic measurements in 11 categories; the collated measurements for an individual, Bertillon maintained, constituted a unique "signature". Where bertillonage had value proved to be not in identifying criminal types but in keeping track of those previously arrested, although it was a cumbersome and highly fallible method – fallible because, obviously, none of these measurements could be made with any great degree of precision. Even so, its underlying notion led to the concept of fingerprinting, developed independently for detection purposes by the Scottish physician

Henry Faulds (1843–1930) in 1890 – who was able to use it to help police solve a crime while he was working in a Tokyo hospital, but whose discovery was ignored back home in the UK – and the English eugenicist Sir Francis Galton (1822–1911) in 1892. Fingerprinting has of course proved a genuinely useful detective tool.

In a paper called "Ugly Criminals", presented at the 2006 American Economic Association Annual Meeting, H. Naci Mocan of the University of Colorado at Boulder and Erdal Tekin of Georgia State University demonstrated that in fact there *is* a correlation between physical plainness and criminality, although this is nothing inherent in the individual but due more to that structural defect in society whereby, irrationally, physical beauty is regarded as a virtue. Plainer folk, lacking this unfair advantage, are statistically more likely to choose criminality as an option.* This is not, however, what Lombroso was thinking about.

Animal Magnetism

Hypnosis was born from the idea of animal magnetism, formulated by Franz Anton Mesmer (1734–1815). Mesmer's conclusions were in turn based on the work of Father Maximilian Hell (1720–1792), an astrologer at the court of Maria Theresa, who had claimed some patients could be cured by strapping magnets to appropriate parts of their bodies. Hell's work was, again in turn, based on that of Paracelsus, who had believed in fitting out patients with all sorts of materials in order to cure specific ailments. So the idea was by no means a fresh one: like so many Great Truths, it had been the object of both attention and scorn for some considerable while.

Mesmer's work might have been accepted much earlier had he been less of a natural showman and charlatan. On arrival, his patients were confronted by a large covered oval tub

* A different statistical study shows that names are important in this respect too: kids with names indicative of lower social standing, whatever their actual background, are less likely to receive the teacher's attention.

containing iron filings, powdered glass, water and bottles – in which latter were propped upright jointed iron rods; these rods projected through holes in the lid of the tub. The patient stood there, holding a rod, until Mesmer himself appeared, dressed much like a fairytale wizard and clutching an iron wand: this he rubbed on the sufferers' affected parts while staring deeply into their eyes.

Sometimes the cure worked, and this Mesmer attributed to the existence and manipulation of a "substance" he dubbed animal magnetism. The "magnetism" component of this probably arose from astrological considerations (Hell, remember, was an astrologer). If stars and planets influenced people, then presumably there was some force which extended across the gulf of space between them and us. It was known that the Earth has a magnetic field, and so it was thought likely the celestial bodies had magnetic fields, too. Could this interplanetary force, then, be magnetism? If so, human beings had to contain something which responded to the magnetic dictates of the stars – i.e., animal magnetism.

Mesmer considered that he himself had an overabundance of animal magnetism, which he could pour in all its healing glory into other people by the touch of his iron wand. His tub was, of course, a reservoir of animal magnetism placed there ready to boost his own personal efforts.

Louis XVI (1754–1793) set up a commission to investigate. The commission found that, say, a subject would respond to the proximity of the mesmerist's hand only if s/he knew (or *thought* s/he knew) that it was there: blindfolded people would not respond to the hand if told it was not there, but would respond if told it was – even if it was not! One person who had in the past tended to go into convulsions under mesmeric influence did so when brought near to a tree which, he was told, had been "magnetized", whether or not a mesmerist had been anywhere near it.

This was fascinating. The phenomenon cried out for further study. But the commission reported only that animal magnetism was a product of overimagination, and left it at that. Of course, in a way their verdict was right – but also horrendously wrong. Quite how much so was revealed in 1841 by Dr James Braid (*c*1795–1860) who, initially sceptical, showed that the effects associated with animal magnetism could be reproduced without any "magnetic" intervention: all that was needed was a fixed stare and a confident manner. Of course, what he had done was identify the phenomenon of suggestion. It was he who called the process hypnotism.

Mary Baker Eddy (1821–1910) was one remarkably late believer in animal magnetism. Not only did she believe that her husband had been murdered by use of a "mesmeric poison", she insisted, when travelling by train, that an additional engine run on ahead in order to drive away any "malicious animal magnetism" that might be lurking in the rails.

Cryonics

Some terminology: *Cryogenics* is the science of the extremely cold, often dealing with temperatures only a few degrees above Absolute Zero (about −273°C, the temperature below which, for good reasons of physics, there can be no colder); *cryobiology* is that part of cryogenics which concerns itself with the effects of intense cold on biological structures. *Cryonics* is a mutant technological offspring of these, dealing with the use of extremely cold temperatures in conjunction with other, rather grisly procedures in order – at least in theory – to preserve dead bodies, or parts of bodies, until such time as medical science has advanced sufficiently to revive them.

The notion flourished in the latter part of the 20th century, but as early as 1773 Benjamin Franklin (1706–1790) had toyed with a very similar fantasy, although "in all probability, we live in a century too little advanced, and too near the infancy of science, to see such an art brought in our time to perfection." By the end of the 19th century the idea that freezing, rather than Franklin's "embalming", might be used to effect a suspension of animation was gaining force thanks to the leaps and bounds that all forms of technology were taking.

The heyday of cryonics began with the publication in 1966 – at approximately the same time as the term "cryonics" was itself coined, by one Karl Werner – of R.C.W. Ettinger's best-seller *The Prospect of Immortality*. In this book was spelled out the seemingly very real possibility that technology would soon be able to freeze or otherwise put into suspended animation the terminally ill such that they could be preserved in a condition of stasis until medical science could cure them. There is nothing inherently silly about this idea except that it relies on technology we do not possess.

Cryonics was rapidly adopted by those who were prepared to believe in any form of magic, but preferably a technological one, if it would offer them the hope of not dying. It almost immediately became an established folk myth that the corpse of Walt Disney (1901–1966) had on his death in 1966 been frozen and planted in the bowels of some Disneyland or other. In fact Disney wasn't frozen, but three decades later, in 1996, Timothy Leary (1920–1996), the one-time philosophical guru of the LSD generation, was less sceptical than Uncle Walt and until the very last moment, when he was dissuaded by a friend, fully intended that his corpse be swiftly handed over to the cryonics company Cryocare.

Perhaps 100 other people, however, have not like Leary balked at the last hurdle; and in one form or another their corpses – or at least their heads – are being carefully tended in a vault somewhere. This figure is actually astonishingly low because, according to a count done by Alex Heard for his entertaining book on American End-Timers, *Apocalypse Pretty Soon* (1999), there are at least 33 establishments offering cryonics services in the USA at the moment. Heard suspects that a good percentage of these firms may be moribund, companies in name only; and this suspicion was backed up by my own experience trawling the internet in search of them, when time and again I found websites that had received no maintenance since the mid-1990s or earlier.

Current cryonics facilities offer two options to their clients: preservation of the whole body or preservation of merely the head, containing the all-important brain. The advantage of the latter option (neurosuspension) is primarily one of cost – the necessary surgical procedures and other preparations prior to freezing cost roughly the same whichever option is chosen, but it's much less expensive to preserve just a head than to keep the entire body. One assumes the notion is that, upon revival, brain transplantation will be a run-of-the-mill procedure . . . and who would want to be stuck in a wrinkly old body for the rest of eternity when there are bound to be beautiful cloned bodies available for the asking?

Neurosuspension's necessary process, decapitation, has led the cryonicists to at least one nasty encounter with the law – indeed, the law has tended to look somewhat askance at the whole business. In December 1987 an elderly woman in California called Dora Kent died and, with her full prior consent, was then decapitated and her head preserved. Unfortunately, this was done with such due diligence that, by the time of the decapitation, although she was *clinically* dead she was not yet *legally* so. It was suspected the head was in the care of the cryonics company Alcor, and the police and coroner's deputies made several raids on its offices, at one stage arresting the company's officers (all charges were almost immediately dropped). The company survived years of wrangling before escaping from the clutches of the legal system, with only defence bills and repair bills (equipment was damaged in police custody) to show for it; luckily none of its existing frozen corpses were thawed or otherwise harmed. Another company was much less lucky in 1981 when a power failure caused a number of bodies to thaw out; the company vanished in a medusan tangle of vengeful lawsuits.

Whether the individual has chosen full or head-only preservation, the treatment of the remains is not merely a matter of sticking them into some glorified fridge. The first stage of the preparation is much like the cardiopulmonary respiration techniques used in hospitals, and has a similar aim: to keep the blood oxygenated and flowing so the cells may likewise remain oxygenated. At the same time, gradual cooling is introduced and various chemicals are injected, all the while the corpse being transported as rapidly as possible to the cryonics facility. Here the process of *perfusion* can begin: essentially this means the corpse's blood is sucked out to be replaced with "cryoprotective" solutions, whose task is to take over the role of blood in oxygen transportation around a body at temperatures far below that at which blood freezes. Once this has been done, the long, slow, rigorously regulated process of chilling can properly begin, until the corpse is at a temperature of about –79°C. Finally, stabilized here, it is placed in its last resting chamber over liquid nitrogen to attain and maintain a temperature of –196°C.

All of this sounds very reassuringly scientific. The difficulty is that the fluid within the individual cells cannot be easily replaced in the way that the blood can, and as the temperature drops this fluid freezes to form ice crystals that puncture the cell wall, destroying the cell's integrity. It is not quite true that trying to revive a frozen person is like trying to turn a hamburger back into a cow, as cryobiologist Peter Mazur once notoriously remarked, but one appreciates the problem he was driving at.

Here the hopeful cryonicist must look towards another branch of future-tech for salvation: nanotechnology. While it would ordinarily be impossibly difficult to repair millions of individual damaged cells, the task would surely be child's play for the similar millions of submicroscopic bodily repair engines envisioned by the nanotechnologists. However, there are no immediate signs of nanotechnology developing this sort of capability in the near future.

An alternative possibility in which the cryonics fraternity sets much store is called *vitrification*. This was developed in the 1980s for the purposes of maintaining organ banks and repositories of other, simpler biological structures. Effectively, vitrification allows liquids to be chilled to a glassy condition without passing through an intervening crystallization phase, so avoiding the shredding of the cell walls. The process is not generally available to the cryonics facilities for the very simple reason that it is both infernally difficult and infernally expensive: no cryonics facility has the requisite cash for the research. Since the organ-bankers aren't doing the research either –

instead placing their faith in cheaper and, they believe, better methods of attaining their objectives – it is unlikely that vitrification will ever achieve the level of sophistication required for it to be applied to whole-body or even just head-only preservation.

The reason for cryonics' fall from grace can be summed up in a nutshell. There are two stages to the whole cryonics venture: (a) the preparation, chilling and preservation, and (b) the reanimation. Cryonicists claim at least limited success with (a). But even they have to admit that they cannot so much as start to guess about what will be required for (b).

THANATOLOGY

The study of death and dying is not in itself a discarded science, but many of its currently accepted bases seem in danger of becoming so. The undisputed pioneer in the field was the Swiss-born Elisabeth Kübler-Ross (1926–2004), who began as a volunteer carer for Holocaust victims, gained a medical degree, and came to the US where, in the early 1960s, she became in effect an advocate for the dying, very creditably persuading her fellow-professionals to start recognizing that the dying, too, had needs that should be attended to. Kübler-Ross claimed there were five stages through which all people go as dying becomes imminent: denial, anger, the attempt to bargain for more time, depression and finally "positive submission" (i.e., acceptance). This scheme has been widely accepted by the medical profession and indeed by the public at large, but in fact there's no real scientific evidence that the dying inevitably go through all these stages or in that order: with the scheme as a preconception, it's all too easy for observers simply to persuade themselves that this is what's going on. The only person who could tell for sure is soon, of course, no longer around to ask.

Or is that the case? In later life, from about 1980 onward, Kübler-Ross began to come out with some increasingly bizarre ideas, not least among them that there is no such thing as death, that it is really the fountain of youth: "People after death become complete again. The blind can see, the deaf can hear, cripples are no longer crippled after all their vital signs have ceased to exist." One can recognize that there's a sort of poetical and/or metaphysical truth behind this, but unfortunately Kübler-Ross began taking it literally, and soon was communicating with what she called "afterlife entities". She began operating workshops at Shanti Nilaya, her "healing centre" near Escondido, California, where the bereaved could communicate with the "afterlife entities" corresponding to their departed loved ones; the distinction between these workshops and seances is a fine one.

All this strangeness culminated in scandal. Among the attractions at Shanti Nilaya were the "healers and spiritualists" Jay and Marti Barham. Thanks to channeling by the Barhams, "afterlife entities" were able to materialize into human form, not just for the purposes of communicating with the bereaved but also, in the case of grieving widows, even of making love to them. When a number of the women came down with the same vaginal infection, suspicions were aroused. When one of the women turned on the light so everyone could see her dear departed better and they instead found Jay Barham naked except for a turban, suspicions intensified, despite his claim that the "afterlife entities" had been able to draw molecules from his body so as to clone a precise replica of it. The last straw was when the cops were called to investigate the sexual molestation of a 10-year-old child and declined to believe the molester had been an "afterlife entity".

Bibliography and Further Reading

Adams, Fred: *Our Living Multiverse: A Book of Genesis in 0+7 Chapters*, New York, Pi Press, 2004

Asimov, Isaac: *Eyes on the Universe: A History of the Telescope*, London, Deutsch, 1976

Bartholomew, Robert E., and Radford, Benjamin: *Hoaxes, Myths, and Manias: Why We Need Critical Thinking*, Amherst, NY, Prometheus, 2003

Bernal, J.D.: *The Extension of Man: A History of Physics Before 1900*, London, Weidenfeld & Nicolson, 1972

Bernasconi, Robert (ed): *American Theories of Polygenesis*, Bristol, Thoemmes, 2002

Bloom, Harold: *Omens of Millennium: The Gnosis of Angels, Dreams, and Resurrection*, New York, Riverhead, 1996

Boorstin, Daniel J.: *The Discoverers: A History of Man's Search to Know his World and Himself*, New York, Random, 1983

Broad, William, and Wade, Nicholas: *Betrayers of the Truth: Fraud and Deceit in the Halls of Science*, New York, Simon & Schuster, 1982

Burger, William C.: *Perfect Planet, Clever Species*, Amherst, NY, Prometheus, 2003

Bynum, W.E., Browne, E.J., and Porter, Roy: *Macmillan Dictionary of the History of Science*, London, Macmillan, 1983

Chapman, Matthew: *Trials of the Monkey: An Accidental Memoir*, London, Duckworth, 2001

Clancy, Susan: *Abducted: How People Come to Believe They Were Kidnapped by Aliens*, Cambridge, Massachusetts, Harvard University Press, 2005

Close, Frank: *Too Hot to Handle: The Race for Cold Fusion*, London, W.H. Allen, 1990

Cohen, I. Bernard: *Revolution in Science*, Cambridge, Massachusetts, Belknap Press/Harvard University Press, 1985

Cohen, J.M. (trans. and ed.): *The Discovery and Conquest of Peru*, revised edition, London, Folio Society, 1981

Collins, Paul: *Banyard's Folly: Thirteen Tales of Renowned Obscurity, Famous Anonymity, and Rotten Luck*, New York, Picador, 2001

Costello, Peter: *The Magic Zoo: The Natural History of Fabulous Animals*, London, Sphere, 1979

Cunnane, Stephen C.: *Survival of the Fattest: The Key to Human Brain Evolution*, Singapore, World Scientific Publishing, 2005

Davies, Gordon L.: *The Earth in Decay: A History of British Geomorphology 1578–1878*, London, Macdonald, nd. but *c*1968

Davies, Paul: *The Fifth Miracle: The Search for the Origin and Meaning of Life*, New York, Simon & Schuster, 1999

Dennett, Daniel: *Darwin's Dangerous Idea: Evolution and the Meanings of Life*, New York, Simon & Schuster, 1995
Dingle, Herbert: *Science at the Crossroads*, London, Martin Brian & O'Keefe, 1972
Drury, Stephen: *Stepping Stones: The Making of Our Home World*, Oxford, OUP, 1999
Dwyer, William M.: *What Everyone Knew About Sex*, London, Macdonald, 1973
Dyson, Freeman J.: *Disturbing the Universe*, London and New York, Harper & Row, 1979
Erickson, George A.: *Time Traveling with Science and the Saints*, Amherst, NY, Prometheus, 2003
Evans, Bergen: *The Natural History of Nonsense*, London, Michael Joseph, 1947
Evans, Bergen: *The Spoor of Spooks*, London, Michael Joseph, 1955
Evans, Christopher: *Cults of Unreason*, London, Harrap, 1973
Feder, Kenneth L.: *Frauds, Myths, and Mysteries: Science and Pseudoscience in Archaeology*, 3rd edn, Mountain View, California, Mayfield, 1999
Fernández-Armesto, Felipe: *Ideas that Changed the World*, New York, Dorling Kindersley, 2003
Ford, Brian J.: *The Revealing Lens: Mankind and the Microscope*, London, Harrap, 1973
Forrest, Barbara, and Gross, Paul R.: *Creationism's Trojan Horse: The Wedge of Intelligent Design*, Oxford, OUP, 2004
Gallant, Rene: *Bombarded Earth*, London, John Baker, 1964
Gardner, James N.: *Biocosm: The New Scientific Theory of Evolution: Intelligent Life is the Architect of the Universe*, Makawan, Hawaii, Inner Ocean, 2003
Gardner, Martin: *Fads and Fallacies in the Name of Science*, New York, Dover, 1952; revised and expanded edition, 1957
Gardner, Martin: *Science: Good, Bad and Bogus*, Amherst, NY, Prometheus, 1989
Garrison, Fielding H.: *An Introduction to the History of Medicine*, 4th edition, Philadelphia and London, W.B. Saunders, 1929
Goldsmith, Donald (ed.): *Scientists Confront Velikovsky*, Ithaca and London, Cornell University Press, 1977
Goldsmith, Donald, and Owen, Tobias: *The Search for Life in the Universe*, Menlo Park, Benjamin/Cummings, 1980
Gordon, Henry: *Extrasensory Deception*, Buffalo, NY, Prometheus, 1987
Gould, Stephen Jay: *Bully for Brontosaurus: Reflections in Natural History*, New York, Norton, 1991
Gould, Stephen Jay: *The Mismeasure of Man*, New York, Norton, 1981

Gregory, Richard L. (ed): *The Oxford Companion to the Mind*, Oxford, OUP, 1987

Gruber, Howard E.: *Darwin on Man: A Psychological Study of Human Creativity*, London, Wildwood, 1974

Hardy, Sir Alister: *The Living Stream: A Restatement of Evolution Theory and its Relation to the Spirit of Man*, London, Collins, 1965

Heard, Alex: *Apocalypse Pretty Soon: Travels in End-Time America*, New York, Norton, 1999

Hines, Terence: *Pseudoscience and the Paranormal*, 2nd edn, Amherst, NY, Prometheus, 2003

Inglis, Brian: *A History of Medicine*, Cleveland, Ohio, World, 1965

Ingram, Jay: *The Barmaid's Brain, and Other Strange Tales from Science*, New York, W.H. Freeman, 1998

Irvine, William: *Apes, Angels, and Victorians*, New York, McGraw–Hill, 1955

Jaki, Stanley L.: *The Milky Way: An Elusive Road for Science*, Newton Abbot, David & Charles, 1979

John, Brian (ed.): *The Winters of the World: Earth Under the Ice Ages*, Newton Abbot, David & Charles, 1979

Keller, Werner: *The Bible as History*, translated by William Neil, London, Hodder & Stoughton, 1956

Khomeini, Ayatollah; *Sayings of the Ayatollah Khomeini* (ed. Jean-Marie Xaviere; English edn. trans. Harold J. Salemson and Tony Hendra), New York, Bantam, 1980

Kitcher, Philip: *Abusing Science: The Case Against Creationism*, Cambridge (Mass.), MIT Press, 1982

Knight, David: *The Nature of Science: The History of Science in Western Culture Since 1600*, London, Deutsch, 1976

Koestler, Arthur: *The Case of the Midwife Toad*, London, Hutchinson, 1971

Koestler, Arthur: *The Sleepwalkers*, London, Hutchinson, 1959

Leakey, Richard E., and Lewin, Roger: *Origins: What New Discoveries Reveal about the Emergence of Our Species and Its Possible Future*, New York, Dutton, 1977

Lerner, Eric J.: *The Big Bang Never Happened*, London, Simon & Schuster, 1992

Lewis, James R. (ed): *Encyclopedic Sourcebook of UFO Religions*, Amherst, NY, Prometheus, 2003

Ley, Willy: *Dawn of Zoology*, Englewood Cliffs, NJ, Prentice–Hall, 1968

Lindsay, Jack: *Blastpower and Ballistics: Concepts of Force and Energy in the Ancient World*, London, Muller, 1974

Lindsay, Jack: *The Origins of Alchemy in Graeco-Roman Egypt*, London, Muller, 1970

Lindsay, Jack: *The Origins of Astrology*, London, Muller, 1971

Luce, J.V.: *The End of Atlantis*, London, Thames & Hudson, 1969
Ludwig, Jan (ed): *Philosophy and Parapsychology*, Buffalo, NY, Prometheus, 1978
Martin, Brian: *Information Liberation: Challenging the Corruptions of Information Power*, London, Freedom Press, 1998
Menzel, Donald H., and Taves, Ernest H.: *The UFO Enigma: The Definitive Explanation of the UFO Phenomenon*, Garden City, Doubleday, 1977
Millar, Ronald: *The Piltdown Men*, London, Gollancz, 1972
Moore, Patrick: *Can You Speak Venusian?: A Guide to the Independent Thinkers*, revised edition, London, Star, 1976
Morgan, Chris, and Langford, David: *Facts and Fallacies: A Book of Definitive Mistakes and Misguided Predictions*, Exeter, Webb & Bower, 1981
Morton, Eric: "Race and Racism in the Works of David Hume", *Journal on African Philosophy*, vol 1, no. 1, 2002
Moseley, James W., and Pflock, Karl T.: *Shockingly Close to the Truth!: Confessions of a Grave-Robbing Ufologist*, Amherst, NY, Prometheus, 2002
Nicholi, Arman M., Jr (ed): *The New Harvard Guide to Psychiatry*, Belknap Press/Harvard University Press, 1988
Numbers, Ronald L.: *The Creationists: The Evolution of Scientific Creationism*, New York, Knopf, 1992
Pais, Abraham: *Inward Bound: Of Matter and Forces in the Physical World*, Oxford, Clarendon Press, 1986
Pallasch, Thomas J., and Wahl, Michael J.: "Focal Infection: New Age or Ancient History?", *Endodontic Topics*, 2003, 4
Pennick, Nigel: *Lost Lands and Sunken Cities*, London, Fortean Tomes, 1987
Pennock, Robert T. (ed): *Intelligent Design Creationism and Its Critics: Philosophical, Theological, and Scientific Perspectives*, Cambridge, Massachusetts, MIT Press, 2001
Penrose, Roger: *The Emperor's New Mind: Concerning Computers, Minds, and the Laws of Physics*, Oxford, OUP, 1989
People for the American Way: *A Right Wing and a Prayer: The Religious Right and Your Public Schools*, Washington DC, People for the American Way, 1997
Plait, Philip: *Bad Astronomy: Misconceptions and Misuses Revealed, from Astrology to the Moon Landing "Hoax"*, Hoboken, NJ, Wiley, 2002
Porter, Roy: *The Making of Geology: Earth Science in Britain 1660–1815*, Cambridge University Press, 1977

Porter, Roy, and Hall, Lesley: *The Facts of Life: The Creation of Sexual Knowledge in Britain 1650–1950*, New Haven, Connecticut, Yale University Press, 1995

Regal, Brian: *Human Evolution: A Guide to the Debates*, Santa Barbara, ABC–CLIO, 2004

Renfrew, Colin: *Before Civilization: The Radiocarbon Revolution and Prehistoric Europe*, London, Cape, 1973

Sagan, Carl: *Broca's Brain: The Romance of Science*, London, Hodder & Stoughton, 1979

Sagan, Carl: *Billions & Billions: Thoughts on Life and Death at the Brink of the Millennium*, New York, Random, 1997

Sagan, Carl: *The Demon-Haunted World: Science as a Candle in the Dark*, London, Headline, 1996

Sagan, Carl: *Pale Blue Dot: A Vision of the Human Future in Space*, New York, Random, 1994

Scull, Andrew: *Madhouse: A Tragic Tale of Megalomania and Modern Medicine*, New Haven, Connecticut, Yale University Press, 2005

Sheehan, Helena: *Marxism and the Philosophy of Science: A Critical History*, 2nd edn, Humanities Press International, 1993

Shermer, Michael: *Why People Believe Weird Things: Pseudo-Science, Superstition, and Bogus Notions of Our Time*, New York, Freeman, 1997

Sladek, John: *The New Apocrypha: A Guide to Strange Sciences and Occult Beliefs*, St Albans, Hart-Davis, MacGibbon, 1974

Smith, Adam: *Powers of Mind*, New York, Random, 1975

Stiebing, William H., Jr: *Ancient Astronauts, Cosmic Collisions, and Other Popular Theories About Man's Past*, Buffalo, NY, Prometheus, 1984

Story, Ronald: *The Space-Gods Revealed: A Close Look at the Theories of Erich von Däniken*, London, New English Library, 1976

Tabori, Paul: *The Natural Science of Stupidity*, Philadelphia, Chilton, 1959; reissued as *The Natural History of Stupidity*, New York, Barnes & Noble Books, 1993

Tannahill, Reay: *Flesh and Blood: A History of the Cannibal Complex*, London, Hamish Hamilton, 1975

Tarling, D.H., and Tarling, M.P.: *Continental Drift: A Study of the Earth's Moving Surface*, London, Bell, 1971

Taylor, F. Sherwood: *The Alchemists: The Founding of Modern Chemistry*, London, Heinemann, 1952

Tipler, Frank J.: *The Physics of Immortality: Modern Cosmology, God and the Resurrection of the Dead*, New York, Doubleday, 1994

Tompkins, Peter: *Secrets of the Great Pyramid*, New York, Harper & Row, 1971

Union of Concerned Scientists: "Scientific Integrity in Policymaking: An Investigation into the Bush Administration's Misuse of Science", Cambridge, Massachusetts, March 2004

Union of Concerned Scientists: "Scientific Integrity in Policymaking: Further Investigation of the Bush Administration's Misuse of Science", Cambridge, Massachusetts, July 2004

Vorzimmer, Peter J.: *Charles Darwin: The Years of Controversy - The* Origin of Species *and its Critics 1859–82*, University of London Press, 1972

Wanjek, Christopher: *Bad Medicine: Misconceptions and Misuses Revealed, from Distance Healing to Vitamin O*, Hoboken, NJ, Wiley, 2003

Ward, Philip: *A Dictionary of Common Fallacies*, Cambridge, Oleander, 1978

Warshovsky, Fred: *Doomsday: The Science of Catastrophe*, London, Abacus, 1979

Wellard, James: *The Search for Lost Worlds*, London, Pan, 1975

Welles, James F.: *The Story of Stupidity: A History of Western Idiocy from the Days of Greece to the Moment You Saw this Book*, Greenport, NY, Mount Pleasant Press, 1988

Whitaker, Robert: *Mad in America: Bad Science, Bad Medicine, and the Enduring Mistreatment of the Mentally Ill*, Cambridge, Massachusetts, Perseus, 2002

White, Michael: *Weird Science: An Expert Explains Ghosts, Voodoo, the UFO Conspiracy, and Other Paranormal Phenomena*, New York, Avon, 1999

Whitcomb, John C. Jr, and Morris, Henry: *The Genesis Flood*, Grand Rapids, Baker, 1961

Williams, William F. (ed.): *Encyclopedia of Pseudoscience*, New York, Facts on File, 2000

Wilson, Edward O.: *Consilience: The Unity of Knowledge*, New York, Knopf, 1998

Wynn, Charles M., and Wiggins, Arthur: *Quantum Leaps in the Wrong Direction: Where Real Science Ends . . . and Pseudoscience Begins*, Washington, DC, Joseph Henry Press, 2001

Young, James Harvey: *The Medical Messiahs: A Social History of Health Quackery in Twentieth-Century America*, 2nd edn, Princeton, NJ, Princeton University Press, 1992

Zangger, Eberhard: *The Flood from Heaven: Deciphering the Atlantis Legend*, London, Sidgwick & Jackson, 1992

INDEX